Integrated Energy Systems with a Carnot Battery

Integrated Energy Systems with a Carnot Battery: A Pathway to Low-Carbon Energy Systems explores Carnot battery technology and its integration into modern energy systems, demonstrating how it can efficiently store and convert energy. It presents optimization approaches of Carnot batteries, including various system layouts, thermal energy storage options, and round-trip efficiency improvements.

Covering the modeling of integrated energy systems (IES), the book focuses on the interaction among various energy carriers (electricity, steam, heat, etc.). It explores the integration of Carnot batteries into IES, featuring case studies illustrating their applications in optimizing system performance, stability, and cost-effectiveness in industrial environments.

This book serves as a comprehensive guide for energy researchers, engineers, and industry professionals seeking innovative solutions for energy storage and conversion.

Integrated Energy Systems with a Carnot Battery

A Pathway to Low-Carbon Energy Systems

Edited by

Xiaojie Lin, Jian Song, and Wei Zhong

CRC Press
Taylor & Francis Group
Boca Raton London New York

CRC Press is an imprint of the
Taylor & Francis Group, an **informa** business

First edition published 2026
by CRC Press
2385 NW Executive Center Drive, Suite 320, Boca Raton FL 33431

and by CRC Press
4 Park Square, Milton Park, Abingdon, Oxon, OX14 4RN

CRC Press is an imprint of Taylor & Francis Group, LLC

© 2026 selection and editorial matter, Xiaojie Lin, Jian Song, and Wei Zhong; individual chapters, the contributors

ISBN: 978-1-041-04980-7 (hbk)
ISBN: 978-1-041-04981-4 (pbk)
ISBN: 978-1-003-63082-1 (ebk)

DOI: 10.1201/9781003630821

Typeset in Times
by SPi Technologies India Pvt Ltd (Straive)

Contents

About the Editors

Xiaojie Lin is Associate Professor at the College of Energy Engineering, Zhejiang University. He graduated with a bachelor's degree in Energy and Environmental Systems Engineering from Zhejiang University in 2012 and obtained a PhD degree in Mechanical Engineering from the University of Maryland, USA, in 2017. During his PhD study, he worked under renowned experts in refrigeration and HVAC fields, researching dynamic simulation modeling and demand response optimization control of variable refrigerant flow systems and phase change materials. After obtaining his PhD, he worked at Mitsubishi Electric R&D Center on dynamic simulation and design of energy systems for zero energy buildings in Japan. In 2018, Dr. Lin joined Zhejiang University as a postdoctoral researcher, focusing on theoretical research on multi-complementary smart heating technologies, and received funding from the "International Postdoc Exchange Program" of the Ministry of Human Resources and Social Security. His current research directions are integrated energy system modeling, control theory, and smart heating technologies.

Jian Song is Assistant Professor at the Birmingham Centre for Energy Storage and School of Chemical Engineering, University of Birmingham, UK; Honorary Research Fellow at the Department of Chemical Engineering, Imperial College London; and Managing Editor of Applied Thermal Engineering. He has been acting as the Deputy Group Leader of Clean Energy Processes Laboratory at Imperial College London prior to joining Birmingham in May 2023. He received his PhD Degree from Tsinghua University in 2018 and was awarded the Zijing Scholar Fellowship, Outstanding PhD Thesis by Chinese Society for Internal Combustion Engines, and Wu Zhonghua Outstanding Postgraduate Award. Dr. Jian Song visited the Whittle Laboratory at the University of Cambridge in 2016 and participated in an internship at Mitsubishi Heavy Industries in 2015. His research focuses on clean and sustainable energy technologies, which covers thermal fluids, processes, devices, and systems, with an emphasis on multi-scale integrated design and optimization of energy conversion and storage options, aiming at enhancing energy utilization efficiency across conventional and renewable resources, and contributing to energy harvesting and sustainability.

Wei Zhong is Professor at the College of Energy Engineering, Zhejiang University. He is Deputy Director of the Institute of Thermal Science and Power Systems, Energy Engineering College and Deputy Director of Power Engineering Center, Polytechnic Institute, Zhejiang University. His current research interests include smart energy system, complex thermal hydraulic system modeling and simulation, thermal system engineering, cyber-physical system, model predictive control, artificial intelligence, industrial big data, and other thermal energy engineering/information technologies. He is also interested in optimization and control of heating system in future smart cities, district integrated energy system regulation, circular economy and industrial ecology, power plant (industrial boiler) performance design and simulation, and smart power plant engineering application technology research. He has published more than 30 papers on SCI and EI and has won more than 10 provincial and ministerial-level scientific and technological progress awards. So far, he has undertaken more than 100 major projects, including the National Key R&D Project, the Provincial Natural Science Funding Project, and other enterprise cooperate projects.

Contributors

Wenxuan Guo
Zhejiang University
China

Xiangrui Jin
Zhejiang University
China

Xiaojie Lin
Zhejiang University
China

Xueru Lin
Zhejiang University
China

Yihui Mao
Zhejiang University
China

Jian Song
University of Birmingham
United Kingdom

Peng Sun
BYD
China

Yuxuan Xia
Zhejiang University
China

Jiahao Xu
Zhejiang University
China

Ziying Zhao
Zhejiang University
China

Haonan Zheng
Zhejiang University
China

Wei Zhong
Zhejiang University
China

Introduction to Carnot Battery

1

Xiaojie Lin, Jian Song and Xiangrui Jin

1.1 INTRODUCTION

The global energy landscape is now undergoing a fundamental transformation driven by the rapid deployment of variable renewable energy sources. Solar and wind power installations have reached unprecedented penetration levels in major electricity markets. It also brings new operational challenges because of the intermittent nature of renewable energy [1]. The prevalence of electricity shortages is due to the intermittent nature of renewable energy and poor coordination between fossil-fuel-based electricity and renewable energy [2]. Since the first industrial revolution, global energy consumption has shown an exponential progress trend. Regarding statistics, the primary energy consumption has increased by about 50 times since 1800. The growth reflects not only the total volume of consumption but also the continuous evolution of the consumption structure. However, the large-scale utilization of conventional fossil fuels supports the progress of industrial civilization. It also harms the global environment and brings a series of crises. According to analyses by the IEA, the burning of traditional fossil fuels accounts for almost 75% of global greenhouse gas emissions. It has made the conventional fossil fuels-dominated development mode difficult to continue. So, due to the increasingly dangerous situation of energy safety and ecosystem pressure, the transition of energy structure has been a conclusion of the international society. Since the

DOI: 10.1201/9781003630821-1

1

signing of the 2015 Paris Agreement, every nation has made a carbon neutrality promise. The one after another action makes energy evolution urgent. The core of transition is to decrease the percentage of traditional fossil fuel in the total energy structure, and largely raise the clean energy, especially the percentage of renewable energy, in the consumption structure of global energy. Solar energy, wind power, tidal energy, and biomass energy are examples of renewable energy. They are doing a great job in energy transition due to the characteristics of renewability, cleanliness, and wide distribution.

China is the biggest nation that has the largest energy consumption and renewable energy production. Its policy of energy transition says a lot about the global energy structure. China has promised to peak carbon emissions by 2030 and to realize carbon neutrality by 2060. It emphasizes developing green energy a lot, and due to this development, China can promote consumption upgrading and quickly construct a new grid dominated by renewable energy. Germany, as a pioneer of energy transition, establishes an Easter package plan that furthermore accelerates the expansion of renewable energy. There is also a law called the renewable energy law, which is a sign of the German energy transition. This renewable energy law completes the early development of renewable energy in Germany by online electricity price subsidy. All these actions German government did was to achieve the goal of increasing 80% national electricity production from renewable energy. America has put a renewable energy quota standard in every state. It aims that the electricity sold must contain a certain proportion of renewable energy, despite the federal policy.

Although the global society has come into the conclusion that it is necessary to increase the renewable energy penetration rate, renewable energy, like wind and solar energy, has inherent uncertainty and intermittency. They are producing a challenge that makes the grid difficult in stable operation and planning patterns. At the same time, the rapid fluctuations in the output of renewable energy also pose a threat to the voltage stability of the grid. Especially in a distribution network, the large-scale distributed solar power integration may cause issues such as local voltage over limits and bidirectional power flow. It sets a higher standard for power quality control and relay protection systems. Coming to the operation and scheduling of the grid, a high renewable energy penetration rate makes the balance between the electricity supply and demand more difficult. In addition, due to objective errors in wind or solar power forecasts, the dispatch department of the grid has to deploy more rotating standby and auxiliary service capacity to deal with this uncertainty. It no doubt increases the total operational cost. For infrastructure, the geographical mismatch between renewable energy resource-rich areas and load centers motivates the huge demand for interregional and high-capacity power transmission. In short, the integration with renewable energy is changing the grid from a rigid structure that relies on dispatchable power generation to a flexible system that urgently needs flexibility and intelligent regulation. Based on the above background, energy storage technologies that bridge the intermittent energy supply and the stable electricity demand are becoming more and more crucial.

1.2 OVERVIEW OF ENERGY STORAGE TECHNOLOGIES

1.2.1 Electrochemical storage technology

Energy storage technologies include variable energy storage forms. Each optimizes for specific applications, duration, and constraints. The evolution of energy storage technologies has been driven by different application requirements. They often range from milliseconds to seasons.

To understand where can Carnot battery can fit in the energy transition steps, some evaluation parameters need to be clear. These broad needs have improved the development of energy storage technologies and optimized them for a specific place within the overall storage requirement space. The complexity of the modern electricity market requires variable storage technologies cooperating to face the full range of grid services and temporal requirements challenges.

Electrochemical storage technologies, which mainly consist of lithium-ion batteries often do a good job in applications requiring rapid response and high power density. These systems can transition between charging and discharging modes within milliseconds. It makes them fully suitable for frequency regulation. However, the cost of electrochemical storage creates high economic barriers for those whose applications are beyond approximately four hours. Energy capacity costs remain high because of their expensive active materials and complex manufacturing processes. Lithium-ion battery can achieve compact deployment. This advantage can be useful in land-constrained cities and applications with limited space. In the past decades, the global manufacturing has been through scale expansion and the popularization of automated production. This trend has promoted the reduction of the cost of lithium-ion batteries largely, and greatly increased the economic competitiveness of its application in the grid. Nevertheless, the economy of electrochemical storage is deeply related to the storage duration. It is proven that applications over 4 hours can lead to a cost rise due to the expensive active material and complex manufacturing processes. This disadvantage brings a main barrier to the development of electrochemical storage as long-duration storage technologies. Additionally, the deep dependency on the key raw materials such as lithium, cobalt, and nickel also brings a potential risk to its supply chain stability and future cost.

The rapid response characteristics of electrochemical storage have made it important for modern electricity systems. Frequency regulation services require response times measured in seconds. However, voltage support services always require faster response capabilities. The power electronics integrated with battery systems make precise control of active and reactive power output possible. These provide grid operators with flexible tools to maintain system stability under off-design conditions. The cycling performance of lithium-ion batteries has improved greatly through materials science advances and cell design optimization. High-quality lithium-ion battery systems can achieve thousands of deep discharge cycles while keeping acceptable capacity at the same time. It makes them economically able for daily cycling applications.

1.2.2 Mechanical storage technologies

Mechanical storage technologies, including Pumped Hydroelectric Energy Storage (PHES) and Compressed Air Energy Storage (CAES), offer proven pathways for long-duration storage. However, this approach faces great geographic and geological constraints that constrain deployment flexibility. PHES requires suitable topography with elevation differences and water resources. This disadvantage constrains installations to mountainous or coastal regions. PHES represents the most mature large-scale storage technology, which has over a century of operational experience. Time has demonstrated excellent reliability and longevity. Existing installations have achieved operational lifetimes over 50 years while keeping excellent performance characteristics. The technology offers high round-trip efficiency. It can also provide valuable grid services such as frequency regulation, voltage support, and black start capabilities. The geographic limitations of PHES have motivated the development of alternative mechanical storage approaches that can operate without the specific above requirements. CAES uses underground caverns or constructed pressure vessels to store energy in the form of compressed gas. However, conventional CAES requires natural gas combustion to generate electricity. This process creates CO_2 emissions that reduce environmental benefits and make projects influenced to fuel price fluctuation.

Advanced-Compressed Air Energy Storage (A-CAES) is designed to eliminate fuel requirements by storing compression heat for use during expansion. It can achieve zero-emission operation while improving round-trip efficiency. These systems face engineering challenges related to high-temperature thermal storage and turbomachinery design. However, it offers pathways for large-scale storage deployment without geographic constraints. Commercial demonstrations of A-CAES technology have been done by technical feasibility. At the same time, areas are identified for continued development. Underground CAES faces geological constraints similar to PHES [3]. It requires suitable rock formations or salt caverns capable of containing high-pressure gas. The suitable geological formations constrain deployment options in many regions. Environmental concerns regarding groundwater impacts pose additional development challenges. A-CAES uses constructed pressure vessels, offers greater siting flexibility but faces higher costs.

Thermal storage without an active heat pump offers low-cost energy capacity through abundant storage media. It also faces fundamental thermodynamic limitations that prevent efficient electricity regeneration. Simple resistive heating creates thermal energy at moderate temperatures, which cannot drive efficient electricity generation. Due to this, it results in poor round-trip efficiency and limited economic value for electricity storage applications. These systems find application primarily for district heat rather than electricity storage. It serves for industrial processes, district heating networks, or building climate control systems.

The thermodynamic limitations of passive thermal storage are mainly because of the low exergy generated by resistive heating [4]. The thermal energy created at middle temperatures lacks temperature difference. It is required to drive efficient electricity generation. This poor efficiency performance brings economic challenges for electricity storage applications, whose competitors can achieve round-trip efficiencies over 80%. However, thermal storage systems do well in applications where they provide valuable services. Industrial process heat, district heating, and building climate control are large markets. Thermal storage can provide valuable services. At the same time, it can avoid the efficiency punishment associated with electricity regeneration. The low cost of thermal storage media enables economical installations for these applications, although the limitations in electricity storage applications.

Concentrated Solar Power (CSP) technology has demonstrated integration of thermal storage with electricity generation successfully. It utilizes molten salt storage systems to extend solar power generation into the evening hours. These systems achieve round-trip efficiencies of about 35–45% while providing dispatchable solar power [5]. The operational experience gained from CSP installations has informed thermal storage development for other applications. At the same time, CSP demonstrates the feasibility of high-temperature thermal storage systems.

1.2.3 Chemical storage technologies

Chemical storage technologies, including hydrogen production and synthetic fuel synthesis, can offer theoretically unlimited duration capabilities. But they currently face significant efficiency penalties and capital cost challenges.

The variable energy conversion steps required for electricity-to-hydrogen-to-electricity pathways introduce substantial losses. They can achieve overall round-trip efficiencies typically ranging from 30% to 50%, which depends on system design and operating conditions. The infrastructure requirements for gas dealing and storage bring complexity and more cost. They also create safety considerations that must be addressed by proper design and operation. Hydrogen production through electrolysis has achieved great technical maturity in commercial systems. It has demonstrated excellent reliability and performance. Proton Exchange Membrane (PEM) offers rapid response abilities that are suitable for renewable energy integration. However, alkaline electrolyzers provide lower costs for steady operation. Solid oxide electrolyzers operating at high temperatures can achieve higher efficiencies by integrating waste heat utilization, but longer startup times are required. The infrastructure requirements for hydrogen storage and utilization bring great barriers to widespread for electricity storage. The low volumetric energy density of hydrogen requires either high-pressure gas storage or cryogenic liquid storage. Both involve substantial infrastructure costs and energy penalties. The handling requirements for hydrogen also bring safety considerations related to leak detection, ventilation, and materials. They must be addressed throughout the system design.

Fuel cell technology for electricity regeneration has achieved commercial maturity in specialized applications. It has developed a lot for grid-scale deployment. PEM fuel cells offer excellent efficiency and load-following capabilities. But they also face cost and durability challenges when expanded to grid applications. Solid oxide fuel cells can achieve higher efficiencies and fuel flexibility. But they also require high-temperature operation which makes system integration and maintenance difficult. The potential for seasonal storage and transportation applications has a strong research interest in chemical storage technologies. But long-duration storage applications requiring weeks or months of storage duration, it may justify the efficiency loss related to chemical storage. The ability to transport stored energy over long distances also creates unique value propositions for chemical storage in applications. Geographical separation between energy production and consumption creates transportation challenges. Synthetic fuel production represents an extension of hydrogen storage. It can use existing fuel infrastructure when providing carbon cycle closure through CO_2 utilization. Power-to-X pathways can produce methane or other synthetic fuels. Due to integration with existing energy infrastructure, it can provide long-duration storage capabilities. However, the additional conversion steps are required for synthetic fuel production. It further brings efficiency loss that limits applications to specialized scenarios.

1.3 DEFINITION OF CARNOT BATTERY

In addition to the above technologies, the emergence of the Carnot battery reflects a broader recognition of thermal energy storage. It can overcome the fundamental limitations when coupled with sophisticated thermodynamic cycles. The heat pump technology during energy charging phases transforms thermal energy into storage tanks. This innovation enables thermal storage systems to compete with electrochemical storage applications. Carnot battery offers unique advantages in duration and component coupling capabilities.

During charging, electric energy is used to move the heat from the low-temperature tank to the high-temperature tank. Such a task may be done with a traditional heat pump (HP), an electric heater, or any other technology. Likewise, in discharging, any heat engine technology may be used, ranging from Rankine, Brayton, or different thermodynamic cycles, to thermoelectric generators [6]. Some efforts have been made to push the boundaries of achievable round-trip efficiency. Recent research in dynamics modeling has enabled detailed analysis of thermal fluctuation in packed-bed storage. Opportunities are revealed for improved thermal management and reduced heat losses [7]. The integration of artificial intelligence and machine learning techniques has further enhanced operational optimization. Predictive control algorithms can optimize loops based on weather forecasts, electricity prices, and thermal demand, maximizing economic returns [8]. These advanced control systems enable autonomous operation. It is adapted to variable conditions without human intervention while maintaining optimal performance by varying operating scenarios.

Carnot battery differs from electrochemical storage in operational principles and modes. Lithium-ion batteries contend with electrolyte decomposition, electrode degradation, and complex state-of-charge dependencies. Carnot battery faces distinct engineering challenges related to regenerator and turbomachinery efficiency, thermal front management, and long-term material stability at elevated temperatures. These differences create unique operational characteristics that include influence maintenance requirements, performance degradation patterns, and system lifetime expectations.

The absence of chemical options in the operation of the Carnot battery eliminates many degradation mechanisms that limit electrochemical storage lifetimes. Thermal loops' effects on solid storage materials represent the primary degradation concern. However, these often act as gradual property changes rather than catastrophic failures. The nature of thermal storage media combined with proven turbomachinery technologies adapted from industrial applications. It is suggested that potential operational lifetimes be measured in decades rather than years. The mechanical nature of Carnot battery components enables maintenance and refurbishment approaches. Turbomachinery components can be rebuilt or replaced to use established industrial practices. However, thermal storage media can be refreshed or upgraded without replacing entire systems. This maintainability characteristic contributes to lifecycle cost advantages and operational flexibility. It becomes more valuable for long-term infrastructure investments.

The abundance of suitable thermal storage materials does a good job ensuring no constraints about deployment at the grid scale. Volcanic rocks, industrial ceramics, and salt-based thermal storage media can be sourced locally in most geographic regions, reducing transportation costs and supply chain risks. This material accessibility contrasts with electrochemical storage dependencies on lithium, cobalt, and other strategic materials. The Carnot battery materials also compare to electrochemical technology, particularly regarding end-of-life management and recycling requirements. Thermal storage media consist of inert materials that can be safely reused of without special handling procedures. The absence of toxic or hazardous materials simplifies regulatory compliance while reducing environmental risks throughout the system lifecycle.

The scalability characteristics of the Carnot battery create significant economic advantages for large-scale installations. The cost per unit of a Carnot battery decreases with scale due to material usage efficiencies and bulk purchasing advantages. This trend creates economic incentives that can achieve superior performance and cost characteristics. This behavior, compared with electrochemical storage, can create more favorable scaling economics. The polyfunctionality of the Carnot battery represents another advantage for significant practical implications in integrated energy systems. Unlike single-direction storage technologies, these technologies inherently produce thermal outputs at variable temperature levels during their operation. During charging, the low temperature tank naturally generates temperatures that are lower than the ambient. It is suitable for cooling applications for industrial refrigeration and district cooling networks. During discharging, the heat engine releases waste heat for space heating, domestic hot water, or low-grade industrial processes. Furthermore, the high-temperature tank itself can be directly used for high-temperature process heat requirements. Carnot battery can also serve industrial applications such as chemical processing, metal working, and steam generation. Especially when energy storage systems operate at temperatures higher than conventional thermal storage systems, Carnot battery has a better ability to provide process heat while keeping electricity storage functionality. It creates a unique operational flexibility for industrial integrated energy systems to optimize through multiple energy requirements.

The temperature levels available from Carnot battery systems are the same as industrial process requirements, district heating networks, and buildings. High-temperature outputs can be over 500°C for industrial processes, and middle temperature outputs provided are well matched with district heating systems when operating at 80–120°C. Low-temperature outputs can also provide cooling services. This operational flexibility enables the Carnot battery can participate in electricity markets, district heating networks, and cooling systems at the same time. It establishes a unique value for integrated energy systems to seek optimization. The ability to deliver variable energy forms from a single system can greatly improve investment to enhance project economics. Economic analyses come to a conclusion that the multi-commodity mode improves project returns compared to a single electricity application. The potential of the Carnot battery matches well with industrial decarbonization strategies that require heat and electricity output at the same time. The integration potential can also be used to urban energy systems. District heating and cooling networks can benefit from the thermal outputs. Smart cities are gradually realizing the importance of achieving decarbonization while improving energy system efficiency. Carnot battery provides enabling technology for offering electricity and thermal energy services. Industrial facilities often have heat demands and electricity demands at the same time. Carnot battery offers optimized operation that maximizes utilization of both electricity and thermal outputs.

Carnot battery has a specialized position in this broader energy storage solution. It is distinguished by particular operations. Unlike geographically constrained solutions that require specific topographies or geological formations, the Carnot battery demonstrates notable siting flexibility. Because thermal storage media can be installed within an industrial footprint. This flexibility contrasts greatly with established large-scale storage approaches. It offers abilities for electrochemical technologies that dominate in shorter-duration applications. The siting flexibility of the Carnot battery enables deployment near data centers, renewable generation facilities, or industrial complexes. This advantage allows optimal integration with existing electrical infrastructure. It can also minimize transmission and losses. The ability to site storage resources optimally within electrical networks creates additional value through congestion relief and transmission deferral benefits.

Urban deployment opportunities for Carnot batteries are because of their contained thermal storage systems. It does not require large water reservoirs or geological formations. Industrial sites, decommissioned power plants, or purpose-built facilities can accommodate Carnot battery installations. At the same time, it provides access for electrical transmission infrastructure and potential thermal energy customers. The compatibility with urban and industrial environments is expanded to deployment opportunities.

The thermodynamic architecture specifically addresses limitations observed in conventional thermal storage without a heat pump. By integrating a heat pump in charging, the Carnot battery establishes a temperature difference. It enables efficient electricity recovery while creating optionality for thermal energy utilization. This architectural approach transforms thermal from a single application to a true electricity storage solution with variable forms.

The heat pump integration in charging brings a fundamental innovation. Rather than simply converting electricity to heat, the heat pump makes the temperature difference achievable by given electrical input. This thermodynamic effect allows practical round-trip efficiencies approaching those of electrochemical storage while utilizing abundant, low-cost thermal storage media.

COP achieved by a heat pump directly influences the thermal storage phase of Carnot battery installations. Advanced heat pump designs utilize optimized refrigerants and component configurations. They can easily achieve COP over 4.0 under appropriate operating conditions. This design creates three defining attributes that distinguish the Carnot battery from the storage technologies. First, the long duration enabled by inexpensive thermal storage materials permits economically viable extension to tens of hours without the steep cost escalation observed in electrochemical alternatives. The marginal cost of storage capacity remains low because thermal storage media are a small percent of the total system cost. It creates economic advantages that increase with duration. The cost structure of the Carnot battery creates a fundamental economic advantage for long-duration utilization. Although power conversion equipment represents the largest capital cost component, thermal storage capacity can be expanded to minimal incremental cost by additional storage media. This cost scaling behavior contrasts with electrochemical storage, where energy capacity costs remain high regardless of duration, creating crossover points beyond which Carnot batteries become increasingly cost-competitive. The economic analysis of duration scalability says that the Carnot battery achieves cost parity with lithium-ion systems. Cost advantages increase greatly for longer duration requirements.

Second, the Carnot battery creates variable revenue streams that improve project economics. This multi-form energy operation reduces dependence on electricity price spreads alone. It can provide additional revenue sources that can justify investment even when electricity arbitrage opportunities are limited. The thermal energy sales often command premium prices compared to electricity, particularly in industrial processes. The potential for revenue creates big value for Carnot battery projects by reducing financial risk and improving overall returns. Electricity market volatility can greatly impact the economics of storage projects. It depends on energy arbitrage, while thermal energy markets often provide more stable and predictable revenue streams. The combination of electricity and thermal revenue sources creates a synergistic effect that reduces overall revenue volatility while improving investment attractiveness. Third, no geographical constraints in Carnot battery makes deployment at strategic grid locations near load centers or renewable generation possible. This flexibility allows optimizing storage deployment to maximize grid benefits.

The benefits of grid integration are not limited to simple energy storage. They also include valuable ancillary services that can improve system stability and reliability. The dispatchable power output from the Carnot battery can provide frequency regulation, voltage support, and system inertia. It is becoming increasingly valuable while conventional thermal generation is retired. These ancillary services can often provide additional sources of income while also helping to improve overall system reliability and stability.

This comprehensive position makes the Carnot battery particularly suitable for applications that require long durations, flexibility, and service provision. The technology effectively bridges between short-duration electrochemical storage and geography-dependent mechanical storage. Carnot battery occupies a unique middle ground in the duration-cost-capability spectrum. This ground becomes increasingly valuable while electricity systems require storage solutions. It can respond to short-term grid balancing issues as well as address longer-term renewable energy integration. The market positioning analysis indicates that in most applications. Carnot batteries are complementary to other energy storage technologies rather a single direct. Short-duration applications that require fast response times can still install electrochemical energy storage. Seasonal storage needs may be better addressed through chemical storage technologies. However, the middle duration range is an optimal interval. Carnot battery has unique advantages in cost and operational flexibility.

With the increasing penetration of renewable energy and the evolving requirements for grid flexibility. The application space for middle-duration energy storage continues to expand. Daily load following, renewable energy shifting, and peak capacity provision all fall within the optimal duration range for Carnot battery deployment. The growing demand for these services creates huge market opportunities. They match well with Carnot battery capabilities and economic characteristics. The core of Carnot battery operation lies in two coupled thermodynamic cycles. Both transform energy between electrical and thermal domains through heat pump and heat engine operations. During the charging phase, electricity input drives the heat pump; at the same time, it raises the temperature of the high temperature tank while depressing the working fluid in the low temperature tank. This dual action makes a temperature difference. It can effectively store electrical energy as thermal energy. Thereby Carnot battery creates a high thermal potential that can drive thermal energy to electricity regeneration. During the discharging phase, the temperature gradient between the two tanks drives a heat engine cycle to regenerate electricity. The entire process operates as a closed thermal loop. It does not release heat rejection to the environment. It creates a thermodynamically reversible framework for energy storage. This closed-loop characteristic makes the Carnot battery unique from traditional energy storage technologies.

The reversibility of thermodynamic processes has a fundamental advantage over traditional thermal storage technologies. Traditional thermal storage technologies face irreversible losses at each energy conversion step. It is especially when converting between electrical and thermal energy. A Carnot battery minimizes these losses by operating as close to reversible conditions as practically possible. It maintains high exergy throughout the storage and retrieval process.

In the field of research and development, two main architectural approaches have emerged. Each offers unique advantages depending on requirements and technical constraints. The Brayton cycle utilizes inert gaseous as working fluids that are compressed and expanded through rotating machinery. They follow mature gas turbine principles adjusted for energy storage applications. The working fluid circulates in a closed loop that includes a compressor, expander, heat exchangers, and thermal storage components.

Thermal energy storage of Brayton-type Carnot battery typically occurs in packed-bed regenerators. They're always full of ceramic spheres or crushed rock particles. Thermal fronts propagate longitudinally during cyclic operation. The packed-bed approach offers excellent thermal capacity at a low cost. At the same time, it allows precise thermal stratification control through careful design of particle size, void fraction, and bed geometry. The thermal wave propagation characteristics can be optimized to minimize mixing and maintain clear temperature transitions to preserve thermodynamic quality.

Counterflow regenerator design has been shown to be essential for minimizing exergy destruction by preheating and precooling working fluid streams. This internal heat recovery significantly improves cycle efficiency by reducing the temperature differences between the working fluid and storage tank. It minimizes irreversible heat transfer losses, which would reduce system performance. The effectiveness of the regenerator becomes a key design parameter with a great impact on round-trip efficiency.

Rankine-based Carnot battery employs a vapor-compression heat pump during the charging phase. It transfers thermal energy to molten salt storage systems at high temperatures. The vapor compression cycle has a high thermal lift capability, enabling the creation of high-temperature tank. Thereby it improves the efficiency. The working fluids in these applications need to be carefully selected to optimize performance in the wide temperature ranges. It encounters during operation while maintaining chemical stability and safety. Power regeneration in Carnot battery is achieved through conventional steam or Organic Rankine Cycles. This cycle expands vapor through turbines, making full use of the duration of development in thermal energy generation. This architectural approach largely benefits from experience in solar power thermal storage. However, it faces challenges in high-temperature heat pump development and thermal management. The integration of heat pump and power generation equipment needs careful optimization to achieve efficient operation in both the charging and discharging phases.

The selection between Brayton type and Rankine type depends on many factors, including temperature, efficiency, cost, and integration with existing infrastructure. Brayton-type typically operate at a higher peak temperature, but they can offer simpler working fluid and potentially higher costs for smaller installations. Rankine type usually achieves lower peak temperatures and may offer integration well with industrial thermal networks. But it requires more complex fluid handling systems and potentially higher capital costs.

Among these two storage architectural approaches, there are various storage media options. They can be broadly categorized by thermal storage mechanisms. These mechanisms determine energy density, temperature stability, and cost. Sensible heat storage remains the most common, using materials such as refractory concrete, volcanic basalt, or molten salt formulations. These materials have proven thermal stability, abundant availability, and low cost. At the same time, it requires easy handling systems. The energy density depends on specific heat capacity and suitable temperature ranges. Its implementations can achieve 50–200 kWh/m^3 depending on operating conditions.

Latent heat storage uses phase-change materials to offer theoretical advantages in energy density and stability but it faces practical application challenges in thermal conductivity, containment, and cycle stability. PCMs can achieve higher energy densities by using the latent heat of fusion or vaporization. It can maintain a relatively constant temperature during phase transitions. However, most PCMs have low thermal conductivity, which complicates heat exchanger design and long-term cycle stability.

Thermochemical storage is an emerging field in which reversible chemical reactions store energy at a higher density with small losses. These systems store energy in chemical bonds, then reverse during discharge to regenerate thermal energy. The energy densities can be hugely over above sensible heat storage. It also eliminates thermal losses during the charging phase. However, material and system complexity currently limit their practical application, keeping them in the research field and development.

Material decision has a profound impact on system performance, installation economy, and operational limitations. They make it an important design consideration that varies significantly depending on application requirements. When selecting between sensible heat storage, latent heat storage, and thermochemical storage, trade-offs must be made in terms of energy density, cost, complexity, and technological maturity. Currently, most applications use sensible heat storage due to its proven reliability and low cost. Researches continue on advanced materials that may have superior performance.

The design of the thermal storage system must also consider thermal cycle effects and long-term stability under repeated operation. When materials operate in corrosive environments at high temperatures, they must keep their thermal properties and structural integrity. The selection process requires careful evaluation of material properties, cost, and environmental impacts to determine for specific applications.

1.4 DEVELOPMENT OF THE CARNOT BATTERY

The research on parameters in the Carnot battery reveals the design sensitivity. These studies have indicated that the round-trip is deeply dependent on minimizing irreversible losses throughout the entire thermodynamic cycle. These heat losses especially occur in heat exchangers and turbomachinery. They also found that the heat utilization of the whole Carnot battery system plays an important role in system economy. It emphasized that multi-department coupling is important for Business Feasibility.

The first hardware demonstration especially verifies the control of heat wave in Packed bed regenerators, and creates a double-direction thermodynamic cycle operating protocol. Laboratory-scale experiments have confirmed the theoretical predictions about thermal stratification in packed beds.

The technological evolution of the Carnot battery has gone through different stages. It started from the theoretical idea. It started from the theoretical idea. Now it is moving toward pre-commercial use. This progress is driven by a growing understanding. Long-duration energy storage is a key gap in renewable energy integration plans. Early academic research set up basic thermodynamic models. It also did parametric analyses. These works found critical performance sensitivities. They also identified design trade-offs. This was done for different system setups and operating conditions. These basic studies gave a theoretical foundation. It was for later experimental validation work. Those works tested core principles in labs and at pilot scales.

The theoretical development stage began with a realization. Conventional thermal storage has basic limits. This is in electricity storage applications. The reason is exergy destruction during resistive heating. Researchers found something. Adding a heat pump during charging can make higher-grade thermal energy. This energy is good for efficient electricity regeneration. This led to a conceptual framework. Now we call it Carnot battery technology. Early analytical models studied thermodynamic limits. They looked at different cycle setups. They set theoretical efficiency bounds. They also found key parameters. These parameters affect practical performance.

Parametric studies showed critical design sensitivities. One is the importance of the temperature ratio. This ratio is between hot and cold reservoirs. Others are heat exchanger effectiveness, component efficiency, and thermal loss management. These analyses showed something. Round-trip efficiency depends a lot on minimizing irreversible losses. This happens throughout the thermodynamic cycle. It is especially true for heat exchangers and turbomachinery parts. The studies also found a big impact. Thermal energy use affects the whole system's economics. This points out the importance of sector-coupling abilities. They are needed for commercial success.

Initial hardware tests focused on two things. First, they wanted to validate thermal wave propagation control. This was in packed-bed regenerators. Second, they aimed to set up operational rules. These rules are for bidirectional thermodynamic cycles. Lab-scale experiments proved theoretical predictions. These predictions were about thermal stratification in packed beds. But the experiments also found practical problems. It is hard to keep thermal front integrity during long cycling. These tests showed something. It is possible to get high thermal effectiveness in regenerative heat exchangers. They also found design parameters. These are key for scaled-up use.

Pilot-scale tests dealt with system integration problems. These include control algorithms. The algorithms are for coordinating heat pump and heat engine operation. They also include thermal management during standby time. And they include performance optimization under different load conditions. These efforts found practical challenges. Keeping thermal stratification integrity is one. Managing turbomachinery efficiency in different operating modes is another. These findings later helped improve designs. The operational experience from these tests gave useful insights. They are about maintenance needs, performance degradation patterns, and optimization ways for commercial systems.

Great progress was made in solving critical subsystem limits. These limits once held back system performance and economic viability. High-temperature heat pump development for Rankine setups moved forward. This was through component innovations and working fluid optimization. It made thermal lift abilities closer to theoretical limits. And it kept reliable operation. Research focused on finding working fluids. These fluids need to stay stable at high temperatures. They also need good thermodynamic properties across a wide temperature range.

The old way of bidirectional turbomachinery designs changed. Now we use asymmetric setups. Here, separate compressors and expanders can be optimized on their own. Each is for its specific operating mode. This design change came from a realization. The best design parameters are very different between charging and discharging. Separate components work better in each mode.

They are better than the compromised bidirectional designs. Using separate machinery also makes maintenance easier. It improves reliability by reducing the complexity of each component.

Thermal loss reduction methods have changed. They started with simple insulation. Now they use advanced control algorithms. These algorithms minimize standby losses during long storage periods [2]. Advanced insulation systems use vacuum panels, aerogel materials, and reflective barriers. They greatly reduce parasitic heat losses. Without them, storage efficiency would drop. Control system innovations made predictive thermal management possible. It optimizes storage temperatures based on predicted charge-discharge patterns. This minimizes thermal losses. And it keeps the system ready to operate.

Heat exchanger technology progress was another key area. Innovations in microchannel designs, printed-circuit heat exchangers, and ceramic heat exchanger materials helped. They made higher effectiveness possible. And they reduced approach temperatures. These improvements directly make thermodynamic efficiency better. They do it by reducing irreversible losses during thermal energy transfer. Advanced manufacturing techniques allow for heat exchanger designs. These designs were not practical. They produce some new ways to optimize performance.

These accumulated improvements slowly changed the Carnot battery. It has beenderived from something academics were curious about. Now it is a credible grid-scale storage solution. In suitable applications, it can compete economically with other technologies. Moving from a theoretical idea to a proven technology needed continuous research. This research dealt with basic thermodynamic principles, practical engineering problems, and economic optimization all at the same time.

Nowadays, development work focuses on three strategic areas. These areas address key barriers to wide commercial use. Integrating the system with industrial thermal networks is a promising way. Using waste heat improves round-trip efficiency. It also provides valuable co-products. Industrial facilities often need a lot of thermal energy at different temperature levels. This creates chances for the Carnot battery. It can provide both electricity storage and process heat services at the same time.

This integration method allows symbiotic operation. Industrial waste heat can preheat working fluids. Or it can supplement thermal storage. This improves overall system efficiency. And it reduces industrial energy costs. Chemical processing plants, refineries, steel mills, and other energy-intensive industries are good partners. They can provide steady thermal energy needs. This complements the variable electricity storage needs. These industrial partnerships create more income sources. They improve project economics. And they help with industrial decarbonization goals.

Reusing retired thermal power generation infrastructure has big economic benefits. It uses existing electrical connections, rotating equipment, and operational experience. At the same time, it keeps valuable grid services. Things like inertia and voltage support. Coal-fired and natural gas power plants that are going to be retired often have valuable electrical infrastructure. They have transmission connections and skilled workers. These can be adapted for energy storage use. The conversion process can use existing turbine-generators, electrical switchgear, and cooling systems. It just replaces fossil fuel boilers with thermal storage systems.

This reuse way provides paths for a just transition. It keeps jobs and economic activity in communities. These communities depend on fossil fuel power generation. Workers with experience in thermal power plant operation have skills that apply directly. They can work with Carnot battery operation. This makes workforce transition easier. And it keeps valuable expertise. The kept electrical connections also maintain grid stability services. These services might be lost when plants are retired. This keeps system reliability and eliminates emissions.

Standardizing modular power blocks and thermal storage components aims to cut engineering costs. It also speeds up deployment through designs that can be copied. Developing standardized setups reduces the custom engineering needed for each installation. It also allows manufacturing scale economies through component standardization. Modular ways also make phased deployment easier. Capacity can be expanded step by step. This depends on demand and economic conditions.

The standardization work focuses on finding the best capacity range for modular units. It balances economies of scale with transportation and installation limits. Factory-built modules can have better quality control and cost reduction. This is compared to systems built on-site. And once site preparation is done, they can be deployed quickly. The modular way also makes financing easier. It allows standardized cost models and performance guarantees. These are based on proven designs.

This maturing technology is now at the edge of commercial use. Several demonstration projects have set up operational records. They also provide useful performance data. This data helps with later commercial-scale deployments. These demonstration projects have many uses. They validate the technology, optimize performance, go through regulatory approval processes, and help with market acceptance. The operational experience from these projects gives essential data. It is for financing commercial installations. And it shows regulatory compliance and safety performance.

Early commercial projects are placed in strategic locations. This maximizes learning chances. And it provides valuable grid services that justify the investment. These projects often combine electricity storage with thermal energy services. This optimizes income sources. And it shows the multi-commodity abilities. These abilities make the Carnot battery different from single-purpose alternatives. Performance data from these installations helps improve later projects. It also builds confidence among developers, utilities, and financial institutions.

The economic features of the Carnot battery are very different from electrochemical storage alternatives. This is because of their unique cost structure. It is good for longer-duration applications. Power conversion components include turbomachinery, heat exchangers, and power electronics. They make up most of the upfront capital costs. Thermal storage media are a relatively small extra expense. They are for extending the discharge duration. This creates an economic model. Energy capacity costs increase in a different way compared to battery technologies. This gives competitive advantages. It is for applications that need longer discharge durations. These are beyond what typical batteries can do.

This cost structure leads to better economics. This happens when duration needs go beyond the 4 to 6 hour range. In that range, electrochemical storage becomes too expensive. The extra cost of more storage capacity in Carnot systems stays low. Thermal storage media are abundant and cheap. The power conversion equipment works for longer durations. And its cost does not increase proportionally. This economic profile fits well with grid storage needs. These needs are for renewable integration. They often require 8–24 h or longer discharge abilities.

Economic viability gets much better when projects use multiple income sources. They don't just rely on electricity price arbitrage. A Carnot battery can handle multiple commodities. So it can take part in electricity markets, thermal energy markets, and ancillary services at the same time. This creates different income sources. It reduces financial risk and improves overall returns. Thermal energy sales often have higher prices than electricity. This is especially true for industrial process heat applications. They need high temperatures.

Having different income sources also helps against risks. It protects from electricity market changes and regulatory shifts. These might affect storage economics. Projects with multiple income streams are more financially stable. They have a lower risk. This makes project financing and development easier. Thermal energy income can also have more predictable cash flows. This is compared to electricity arbitrage. Electricity arbitrage depends on changing price differences.

Three main implementation paths have appeared in project development. Each targets different market chances and value propositions. Electricity-centric installations target grid locations. These places have frequent renewable curtailment or transmission congestion. They focus on energy time-shifting and providing capacity value. These projects aim to maximize electricity storage abilities. They might also use thermal outputs for local heating. This is where there are markets for it.

The electricity-centric way is good for regions. These regions have high renewable use. Curtailment here means lost economic value. Storage can capture this value. Putting the installations near renewable generation facilities allows direct integration. This reduces transmission needs. And it provides grid stability services. These installations also provide capacity value during peak demand times. And they support renewable energy integration goals.

Integrated energy parks combine electricity storage with district heating or industrial steam supply. They get extra value from thermal energy sales. And they improve overall system efficiency through working together. These installations optimize across multiple energy types at the same time. They often get better economic returns. This is compared to projects that only deal with electricity. Thermal integration also improves round-trip efficiency. It uses thermal outputs that would otherwise be system losses.

District heating applications are very good opportunities. This is in climates with high heating needs. Thermal energy here has higher prices than electricity. Industrial steam supply offers year-round thermal energy demand. It has higher prices. And it supports industrial decarbonization goals. This integrated way creates win-win situations. Energy storage provides grid services. And industrial or district thermal networks get reliable, clean thermal energy.

Infrastructure repurposing projects turn decommissioned fossil generation sites into zero-emission storage facilities. They keep valuable grid connection abilities. And they eliminate combustion emissions. These projects use existing electrical infrastructure, skilled workers, and community acceptance. They also provide paths for a just transition. This is in communities that depend on fossil fuels. The conversion way can often have better economics. This is compared to new greenfield projects. It uses existing infrastructure investments.

The repurposing path also deals with social and political problems. These are related to power plant closures. It keeps jobs and economic activity. And it helps with decarbonization goals. Communities benefit from retained jobs and economic activity. They contribute to the clean energy transition. They don't just face economic loss from plant closures. This way shows that decarbonization can bring economic chances. It's not just about costs for affected communities.

Each implementation path has unique advantages. This depends on local market structures, regulatory frameworks, and existing infrastructure features. Project developers must carefully study local conditions. They need to find the best way. It should maximize value creation. And it should minimize development risks and costs. The different implementation paths give flexibility. This allows the Carnot battery to be used in different market conditions and regulatory environments.

Successful implementation now depends more on policy frameworks. These frameworks need to recognize the unique value of long-duration storage technologies. They should provide proper compensation. This is for the different services these systems can offer. Market designs are essential for economic viability. They need to properly compensate for capacity provision, grid stability services, and system adequacy contributions. Traditional energy arbitrage alone may not be enough. It may not justify the investment costs for long-duration storage systems.

Current electricity markets often underestimate the full range of services from energy storage. Especially those related to system reliability, grid stability, and support for renewable energy integration. Improved market designs that better recognize and compensate these services create more favorable economic conditions for energy storage deployment. They also boost overall system efficiency. Long-duration storage offers special value during prolonged renewable energy shortages. These shortages threaten system adequacy. But current market mechanisms may not fully compensate for this capability. Regulatory clarity on multi-commodity operations promotes participation across electricity and thermal markets. There are no regulatory barriers that might limit project development [6]. Traditional regulatory frameworks usually separate the electricity and thermal sectors. This creates obstacles for projects spanning both markets. Regulatory adaptations that recognize the benefits of sector coupling while maintaining proper oversight create favorable conditions for integrated energy projects. Technical standards for grid interconnection and performance verification reduce project development risks and financing costs. They do this by providing clear requirements and standardized approval processes. Standardized interconnection procedures cut development time and costs. They also give certainty to project developers and utility companies. Performance standards support standardized financing methods. These methods are based on proven technical performance. There's no need for customized risk assessments for each project. These favorable conditions develop alongside technological maturity. Together, they determine the speed and scale of commercial deployment in different power systems and market environments. The alignment between technical capabilities, market demand, and regulatory frameworks will ultimately shape the role of the Carnot battery in future energy systems. Current trends show a growing recognition of long-duration storage's value proposition. There's also increasing policy support for technologies. These technologies can address renewable energy integration challenges and provide multiple energy services. The combination of technological maturity, market demand, and policy support indicates a promising future for the Carnot battery. They are expected to achieve significant growth over the next decade. Globally, power systems are struggling with renewable energy integration challenges. These challenges require long-duration storage solutions. Carnot battery has unique advantages. They offer duration scalability and sector coupling capabilities. These features fill key gaps in the current energy storage technology portfolio. They also provide economic benefits in appropriate applications.

REFERENCES

[1] P. Penglai, Qibin Li, et al. (2024) Thermo-economic and life cycle assessment of pumped thermal electricity storage systems with integrated solar energy contemplating distinct working fluids. *Energy Conversion and Management, 318*: 118895.

[2] Zhi Zhang, Ming Zhou, et al. (2024) Decarbonizing the power system by co-planning coal-fired power plant transformation and energy storage. *Journal of Energy Storage, 66*: 107442.

[3] Alessio Tafone, Alessandro Romagnoli. (2023) A novel liquid air energy storage system integrated with a cascaded latent heat cold thermal energy storage. *Energy, 281*: 128203.

[4] Yao Zhao, et al. (2022) Thermodynamic investigation of latent-heat stores for pumped-thermal energy storage. *Applied Thermal Engineering, 55*: 105802.

[5] Ali Khosravani, Julia K. Sieving, Blake W. Billings, Kody M. Powell (2025) Techno-economic analysis of long-duration energy storage integrated with high-penetration renewable energy systems. *Energy Reports, 14*: 4086–4110.

[6] IRENA. *Comprehensive Technology Assessment and Market Analysis. Innovation Outlook: Thermal Energy Storage.*

[7] A. J. White et al. (2016) Analysis and Optimisation of Packed-Bed Reservoirs for Electricity Storage Applications. *Journal of Power and Energy, 230*: 739–754.

[8] Yongliang Zhao et al. (2023) Multi-objective thermo-economic optimisation of Joule-Brayton pumped thermal electricity storage systems: Role of working fluids and sensible heat storage materials. *Applied Thermal Engineering, 223*: 119972.

Design and Optimization of Carnot Battery

2

Xiangrui Jin, Yuxuan Xia and Xiaojie Lin

2.1 THERMODYNAMIC MODELING OF CARNOT BATTERY

The accurate modeling of Carnot Battery (CB) systems represents one of the most fundamental and challenging aspects of thermo-mechanical energy storage research. Since Fritz Marguerre first patented the concept of CB over a century ago [1], the field has developed to encompass sophisticated modeling approaches that capture the complex thermodynamic, dynamic, and economic characteristics of these systems. Contemporary CB modeling can be systematically divided into three distinct levels: steady-state modeling for thermodynamic cycle analysis, dynamic modeling for transient behavior characterization, and long-term operational modeling incorporating comprehensive economic evaluation. The thermodynamic modeling of the CB will be introduced from two aspects: steady-state modeling and dynamic modeling.

DOI: 10.1201/9781003630821-2

2.1.1 Steady-state modeling framework

2.1.1.1 Core methodology

Steady-state modeling is the foundational approach in CB research, primarily utilized for system design, performance evaluation, and parametric sensitivity analysis. This method assumes that all system components are in thermodynamic equilibrium conditions. According to leveraging energy-saving principles and thermodynamic cycle analysis, it can quickly evaluate the performance indicators of CB. The fundamental approach relies on the first law of thermodynamics. During this process, the energy balance equation will be systematically applied to each component, ensuring that input energy equals output energy plus irreversible losses.

In the CB modeling process presented by McTigue et al. [2], the system's state points are first determined, including the inlet and outlet of the compressor, heat exchanger, expander, and the high- and low-temperature states of the thermal storage tanks. Each state point is rigorously defined by using appropriate equations of state for the working fluid. Poletto et al. [3] provide a steady-state model of a classical CB and evaluates the energy efficiency of each system component, applying energy conservation to the compressor, expander, evaporator, condenser, and thermal storage sections, forming a complete steady-state energy flow loop. For compressor modeling, energy balance is expressed through enthalpy changes equal to compression work input, with isentropic efficiency parameters accounting for real-world irreversibilities. Similarly, expander modeling incorporates isentropic efficiency considerations during the expansion process, while heat exchanger modeling requires simultaneous energy balance evaluation for both hot and cold fluid streams, accounting for temperature difference losses during heat transfer.

Under the steady-state assumption, thermal storage tank modeling treats the storage system as an ideal thermal reservoir with uniform internal temperature distribution. This approach neglects heat propagation delays, thermal fronts within storage media, pressure drops, heat losses, and thermal capacity gradients. The tank temperature will be input as the boundary condition during the calculation of the heat exchanger. This approach largely simplifies the complexity of the overall modeling and also maintains reasonable accuracy for initial design evaluations.

2.1.1.2 Key performance parameters

The pressure ratio and temperature ratio emerge as critical parameters which determine both thermodynamic state and round-trip efficiency of CB systems. The pressure ratio can directly control the outlet temperature of the compressor and expander, thus establishing high-temperature (HT) and low-temperature (LT) boundary conditions. Its optimization affects not only energy storage density but also cycle thermal efficiency, which requires careful balance between performance enhancement and equipment cost considerations.

Thess [4] provides a detailed theoretical discussion of the central role of pressure and temperature ratios in determining thermal performance. The study points out that the pressure ratio is proportional to the compressor inlet and outlet temperatures, which sets the maximum temperature of the heat pump (HP). Thus, it determines the feasible temperature ratio range of the storage system, significantly impacting round-trip efficiency and energy density. As the pressure ratio increases, system efficiency improves monotonically. However, when the pressure ratio is relatively high, the improvement in efficiency will gradually reach saturation and may cause additional heat loss, increasing equipment costs. The authors further note that under specific heat exchange and device efficiency conditions, there exists an optimal pressure ratio that maximizes the system efficiency.

For the temperature ratio, Thess [4] points out in the theoretical model that a nonlinear relationship exists between round-trip efficiency and the temperature ratio. When it is below a certain threshold (approximately 2.2 to 3.0), efficiency increases significantly with rising temperature ratio, but once this value is exceeded, the efficiency improvement plateaus and the marginal gain diminishes. This behavior establishes critical design guidelines for optimizing the trade-off between energy density and system reliability.

2.1.1.3 Limitations and evolution

Although steady-state modeling plays a crucial role in CB research, it has rather prominent limitations when representing actual operating conditions and control strategy responses. The assumption of zero time derivatives for system parameters neglects charging, discharging, and part-load operations, limiting applicability in control-oriented studies. Additionally, the idealized treatment of thermal storage and cooling components fails to capture dynamic behavior during thermal storage processes; therefore, more sophisticated modeling methods need to be developed.

2.1.2 Dynamic modeling approaches

2.1.2.1 Lumped parameter modeling

The evolution toward dynamic modeling began with lumped parameter methods, treating system components as thermodynamic units with spatially uniform properties. White et al. [5] developed a detailed lumped parameter model for a typical CB system, defining four key state points in the cycle, representing the working fluid after compression, thermal storage, expansion, and cooling. The system's cycle efficiency is expressed as a function of these four temperature points and the isentropic efficiencies of each component. To capture the impact of irreversible processes, the authors introduce two parameters: the isentropic efficiency correction factor α, representing the non-ideal behavior of the compressor and expander, and the thermal loss coefficient ξ, which accounts for energy dissipation during long-term operation. Energy flow balance derives from fundamental control-volume energy conservation equations. Net energy loss is defined by differences among compression work, expansion output, and thermal storage heat input/output.

The primary advantage of lumped parameter modeling is that it is able to reveal intrinsic coupling mechanisms through simplified parameter expressions and control-volume energy balance equations. This approach offers excellent analytical tractability and visualization capability, which is suitable for early stage design, parameter optimization, and performance trend evaluation. However, this method neglects the spatio-temporal distribution and dynamic response characteristics, which limits its applicability for simulating actual system behavior under non-steady scenarios such as startup, switching, or dynamic load response.

2.1.2.2 Distributed parameter modeling

To overcome lumped parameter limitations, researchers have progressively adopted distributed parameter modeling, particularly for thermal storage tanks and heat exchanger simulation. White et al. [6] applied a one-dimensional convection-conduction model for the dynamic modeling of the HT, explicitly introducing spatial coordinates and dividing the tank into multiple discrete segments, with energy conservation equations solved using the finite difference method. This approach can capture key dynamic behaviors such as thermal stratification and thermal front propagation, revealing the relationships between thermal front velocity and temperature gradients under different mass flow rates.

Distributed modeling reveals relationships between thermal front velocity and temperature gradients under varying mass flow rates, providing insights into thermal inertia responses and system stability assessment. Although computational demand increases compared to lumped models, this methodology offers significantly enhanced credibility in simulating realistic thermal behavior and dynamic system responses.

2.1.2.3 System-level coupled modeling

Advanced implementations have progressed to system-level coupled dynamic modeling, integrating components such as compressors, heat exchangers, thermal storage tanks, and expanders into unified dynamic frameworks. Yang et al. [7] developed a complete dynamic model of a CB using a modular subsystem architecture, modeling compressor speed response, thermal stratification in the storage tank, and the transient heat transfer behavior of the heat exchanger, employing a thermodynamic state-point tracking algorithm to achieve system-wide coupled evolution.

2.1.2.3.1 Critical dynamic processes

Thermal stratification within thermal storage tanks represents the most significant dynamic process in CB systems. During charging or discharging operations, heat propagates through storage media as thermal fronts, creating distinct temperature gradients influenced by mass flow rate, velocity distribution, thermal diffusivity, and tank geometry. White et al. [6] constructed a one-dimensional unsteady-state energy conservation equation to model this phenomenon, incorporating convective and conductive terms to successfully reproduce the advancement speed of the thermal front under varying flow velocities and to reveal the regulating effect of stratification thickness on thermal efficiency.

Dynamic heat exchanger response constitutes another critical process, where working fluids undergo rapid heat exchange with heat capacity, flow velocity, and wall thermal resistance jointly determining heat transfer rates and temperature matching behavior. Yang et al. [7] provide a detailed study of this phenomenon and identify thermal lag in the heat exchanger, which can easily lead to temperature instability.

The most complex dynamic processes involve startup, shutdown, and load response characteristics of compressors and expanders. Pecchini et al. [8] employed mechanical performance maps instead of conventional constant isentropic efficiency models to capture the transient output characteristics of devices under varying inlet conditions and rotational speeds. In their simulation, variable inlet guide vane control was applied to limit temperature fluctuations during part-load operation, yet compression lag due to system inertia was still observed. Additionally, Sun et al. [9] demonstrated that during the transition from HP to heat engine mode in a CB system, the sudden temperature shift at the thermal storage tank outlet leads to a sharp rise in instantaneous power, followed by thermal mismatch and efficiency degradation.

2.1.3 Economic modeling framework

2.1.3.1 Cost structure analysis

Techno-economic modeling of CB systems fundamentally aims to translate equipment design and operational parameters into financial evaluation metrics. The modeling framework typically employs life cycle cost (LCC) methodology, with core components comprising capital expenditures (CAPEX) during construction phases and operational expenditures (OPEX) during operational phases.

CAPEX generally encompasses major equipment procurement and manufacturing costs, civil works and installation expenses, system integration and control system costs, grid connection investments, and in high-fidelity models, one-time expenses including commissioning, insurance, and land acquisition. Li et al. [10] divided CAPEX into six items: compressor, expander, reactor, pump system, heat exchanger, and control system, with the first three accounting for more than 70% of total CAPEX, indicating a representative economic model structure.

OPEX covers periodic maintenance requirements, consumable replacement costs, electricity and cooling energy consumption, labor expenses, and potential downtime compensation. The comprehensive treatment of these cost components enables accurate assessment of long-term economic viability and comparison with alternative energy storage technologies.

2.1.3.2 Performance indicators

Economic performance evaluation primarily focuses on two core categories: cost-based metrics such as Levelized Cost of Storage (LCOS) and investment return metrics based on discounted cash flow analysis, including Net Present Value (NPV). LCOS represents the most widely utilized metric, expressing average LCC per unit of electricity output and enabling cross-technology comparisons among different energy storage systems.

The LCOS calculation incorporates total costs in each operational year, including investment amortization, operation and maintenance, replacement, and energy consumption expenses, discounted over the system lifetime. This approach captures compound effects of utilization rate, life cycle duration, and load factor on economic performance.

For investment-return-oriented evaluation frameworks, NPV offers a more investor-centric perspective by assessing absolute economic value under current conditions through discounting and summing net cash flows over the entire project lifetime. Iqbal et al. [11] developed a comparative model between HT and LT PTES systems, noting that although the HT configuration has a lower LCOS, its higher upfront investment and longer commissioning period result in a lower NPV over a 10-year lifetime, suggesting that short-term investors may prefer the LT alternative.

2.1.3.3 Policy impact integration

Policy mechanisms significantly influence investment returns, risk structures, system design, deployment pace, and market positioning of CB systems.

Techno-economic models incorporate three primary policy categories: upfront investment support policies, including capital subsidies and investment tax credits (ITC); operational revenue enhancement policies, such as time-of-use pricing and ancillary service compensation; and externality internalization mechanisms, including carbon pricing systems and renewable energy quotas.

Research demonstrates that well-structured policy frameworks can substantially improve CB economic competitiveness. For example, Shamsi et al. [12] modeled a combined heat and power PTES system under three policy scenarios: no incentives, electricity price support, and ITC. The results showed that with a 30% ITC, the system's LCOS decreased by approximately 18%, and the NPV shifted from negative to positive. Studies from China incorporated capacity pricing, mandatory storage quotas, and time-of-use electricity tariffs into the model simultaneously, finding that when the peak-valley price spread exceeds 0.6 CNY/kWh and the capacity payment surpasses 100 CNY/kW, PTES becomes more economically competitive than electrochemical storage.

2.1.4 Component-level modeling

2.1.4.1 Heat exchanger model

2.1.4.1.1 Design condition

The steady-state model of the evaporator and condenser is constructed based on the first law of thermodynamics and heat transfer equations:

$$h_{steps} = \left[h_{in}, h_{sat\_gas}, h_{sat\_liquid}, h_{out} \right] \tag{2.1}$$

where h_{steps} is defined as enthalpy node values for segmentation (J/kg), h_{sat_gas} is defined as enthalpy of saturated gas phase (J/kg), h_{sat_liquid} is defined as enthalpy of saturated liquid phase (J/kg), h_{in} stands for inlet enthalpy (J/kg), and h_{out} stands for outlet enthalpy (J/kg).

2.1.4.1.2 Off-design condition

When the heat exchanger is operating under off-design conditions, the corrected characteristic line correlation is as follows:

$$\frac{\dot{m}\sqrt{T_{in}}}{p_{in}} = k\sqrt{1 - \left(\frac{p_{out}}{p_{in}}\right)^2} \tag{2.2}$$

2.1.4.2 Compressor model

2.1.4.2.1 Design condition

When compressing gas from the same state to the same pressure, the ratio of the isentropic compressor work to the actual compressor work is defined as the compressor isentropic efficiency, as shown in the following equation:

$$\eta_s = \frac{h_{out,s} - h_{in}}{h_{out} - h_{in}} \tag{2.3}$$

where $h_{out,s}$ is defined as the enthalpy of the working fluid at the compressor outlet for the isentropic compression process (kJ/kg), h_{in} is defined as compressor inlet working fluid enthalpy (kJ/kg), h_{out} is defined as compressor outlet working fluid enthalpy (kJ/kg), and η_s stands for isentropic efficiency.

2.1.4.2.2 Off-design condition

When the compressor operates under off-design conditions, two-dimensional characteristic mapping parameters are applied. The dimensionless parameters for equivalent speed and equivalent mass flow rate are defined as follows:

$$X = \sqrt{\frac{T_{in,design}}{T_{in}}} \tag{2.4}$$

where X represents the equivalent speed, $T_{in,design}$ is the working fluid inlet temperature under design conditions (K), and T_{in} is the actual working fluid inlet temperature under real conditions (K).

$$Y = \frac{\dot{m}_{in} \cdot p_{in,design}}{\dot{m}_{in,design} \cdot p_{in} \cdot X} \tag{2.5}$$

where Y denotes the equivalent mass flow rate, $\dot{m}_{in,design}$ is the working fluid inlet mass flow rate under design conditions (kg/s), $p_{in,design}$ is the working fluid inlet pressure under design conditions (Pa), $\dot{m}_{in}$ is the actual working fluid inlet mass flow rate under real conditions (kg/s), and p_{in} is the actual working fluid inlet pressure under real conditions (Pa). From this, the isentropic efficiency and pressure ratio of the compressor under off-design conditions can be calculated as follows:

$$\eta_s = \eta_{s,design} \cdot g(X,Y) \cdot \left(1 - \frac{igva^2}{10000}\right) \tag{2.6}$$

$$pr = pr_{design} \cdot f(X,Y) \cdot \left(1 - \frac{igva}{100} \right) \tag{2.7}$$

where $igva$ is the inlet guide vane opening (°), $g(X,Y)$ is the efficiency characteristic mapping function, $\eta_{s,design}$ is the compressor design isentropic efficiency, pr_{design} is the compressor design pressure ratio, and $f(X,Y)$ is the characteristic mapping function.

2.1.4.3 Throttle valve model

2.1.4.3.1 Design condition

The core assumption of the throttle valve is that the throttling process involves no energy exchange, and the enthalpy values at the inlet and outlet are the same, as shown in the following equation:

$$h_{out} = h_{in} \tag{2.8}$$

2.1.4.3.2 Off-design condition

Under off-design conditions, the actual pressure loss is related to the fluctuations in the working fluid mass flow rate, as shown in the following equation:

$$\Delta p = p_{in} - p_{out} = f(\dot{m}) \tag{2.9}$$

2.1.4.4 Motor model

To simplify the model, the dynamic response process of the motor under grid frequency regulation is ignored, and its energy conversion efficiency is defined as follows:

$$\eta_{motor} = \frac{P_{shaft}}{P_{elec}} \tag{2.10}$$

2.1.4.5 Thermal storage tank model

The dual-tank thermal storage is the energy storage medium in the thermally integrated CB. It physically separates the HT and LT tanks, enabling the storage of thermal energy from both the HP and Organic Rankine Cycle (ORC) cycles. In other words, the energy storage phase acts as a key connection between the charging and discharging stages, with the thermal energy transfer, mass flow rate, and temperature gradient of the storage medium forming the essential link. The relationship between these three variables is as follows:

$$\dot{Q}_{ch} = \dot{m}_{ch} c_p \left(T_{hot,tank} - T_{cold,tank} \right)$$
$$\dot{Q}_{dis} = \dot{m}_{dis} c_p \left(T_{hot,tank} - T_{cold,tank} \right) \tag{2.11}$$

where $\dot{Q}_{ch}$, $\dot{Q}_{dis}$ represent the heat storage and heat release of the CB under the energy storage and energy release conditions, respectively, (W); $\dot{m}_{ch}$ and $\dot{m}_{dis}$ are the corresponding mass flow rates of the thermal storage medium (kg/s); c_p is the specific heat capacity of the thermal storage medium (kJ/(kg·K)); and $T_{hot,tank}$, $T_{cold,tank}$ are the fluid storage temperatures of the HT and LT thermal storage tanks, respectively.

2.1.4.6 Working Fluid Pump Model

2.1.4.6.1 Design condition

The pump energy conversion efficiency is defined by the isentropic efficiency:

$$P = \frac{\dot{m} \cdot \left(h_{out} - h_{in} \right)}{\eta_p} \tag{2.12}$$

where η_p stands for pump isentropic efficiency.

2.1.4.6.2 Off-design condition

$$\eta_p = \eta_{p,design} \cdot g \left(\frac{\dot{V}}{\dot{V}_{design}} \right) \tag{2.13}$$

2.1.4.7 Turbine model

2.1.4.7.1 Design condition

The ratio of the actual turbine work to the isentropic turbine work is defined as the turbine isentropic efficiency, as shown in the following equation:

$$\eta_s = \frac{h_{out} - h_{in}}{h_{out,s} - h_{in}} \tag{2.14}$$

where η_s is defined as the turbine isentropic efficiency.

2.1.4.7.2 Off-design condition

Similar to the compressor model, the isentropic efficiency of the compressor under off-design conditions can be calculated as follows:

$$\eta_s = \eta_{s,design} \cdot f\left(\frac{\dot{m}}{\dot{m}_{design}}\right) \tag{2.15}$$

2.1.4.8 Motor model

To simplify the model, the dynamic response capability of the generator during grid frequency regulation is ignored, and the energy conversion efficiency of the generator is expressed as follows:

$$\eta_{gen} = \frac{P_{elec}}{P_{shaft}} \tag{2.16}$$

where η_{gen} is defined as motor efficiency, P_{shaft} is motor input shaft power (W), and P_{elec} stands for output electrical power (W).

2.2 SYSTEM LAYOUTS AND COMPONENT SELECTION

2.2.1 System layouts of Carnot battery

The architectural design of CB represents a fundamental departure from conventional electrochemical storage paradigms, embodying a thermodynamic approach to energy storage that leverages established heat engine and HP technologies. The system's conceptual foundation rests upon the reversible thermodynamic principle where electrical energy undergoes a dual conversion process: first from electricity to thermal energy during charging, followed by thermal-to-electrical conversion during discharging. This dual conversion mechanism, while introducing inherent thermodynamic losses, enables unprecedented scalability and longevity characteristics that address the fundamental limitations of electrochemical storage systems. The basic system architecture comprises four primary subsystems: the HP cycle for charging operations, the power generation cycle for discharging, thermal storage reservoirs, and the interconnecting heat exchange network. During charging operations, surplus electrical energy drives a thermodynamic cycle that extracts thermal energy from a LT reservoir and delivers it to a HT reservoir, creating and maintaining a thermal gradient. The discharging process reverses this energy flow, utilizing the stored thermal gradient to drive a power generation cycle that converts the temperature differential back into electrical energy.

The fundamental system layout can be categorized into two primary configurations based on the thermodynamic cycles employed: Rankine-based systems and Brayton-based systems. Each configuration presents distinct advantages and operational characteristics that influence their suitability for specific applications and operating conditions.

2.2.1.1 Rankine-based system layout

The Rankine CB represents the most technologically mature system configuration, leveraging established HP and ORC. The system employs liquid-vapor phase change processes throughout both charging and discharging operations, utilizing the substantial latent heat of vaporization to achieve high energy density storage. The charging cycle typically implements a conventional vapor-compression HP, while the discharging cycle utilizes an ORC optimized for the specific temperature range and working fluid characteristics. The thermal coupling between the HP and ORC cycles occurs through the thermal storage system, which must accommodate the specific temperature and pressure requirements of both cycles. This configuration typically operates at moderate temperatures, generally below 300°C, which aligns well with the critical temperature limitations of organic working fluids and enables the use of conventional materials and components throughout the system. Typical Rankine CB is shown in Figure 2.1.

This figure illustrates an integrated energy storage system combining an ORC with a HP. In the upper ORC cycle, HT thermal storage supplies heat to vaporize the working fluid, which then drives a turbine connected to a generator (G) for power production. In the lower HP cycle, a compressor elevates the working fluid temperature, and heat is released to the HT storage through a condenser before the fluid is throttled and absorbs heat from the LT storage. The two cycles are coupled through the thermal storage tanks, enabling the HP to upgrade and store low-grade heat in HT, which the ORC later extracts to generate electricity, achieving electricity-to-heat-to-electricity energy conversion.

The system layout incorporates several critical components that determine overall performance characteristics. The evaporator on the HP side typically operates at temperatures determined by the available waste heat source or ambient conditions. The condenser operates at elevated temperatures corresponding to the HT thermal storage requirements. On the ORC side, the evaporator extracts thermal energy from the HT storage, while the condenser rejects waste heat to the environment or LT storage system.

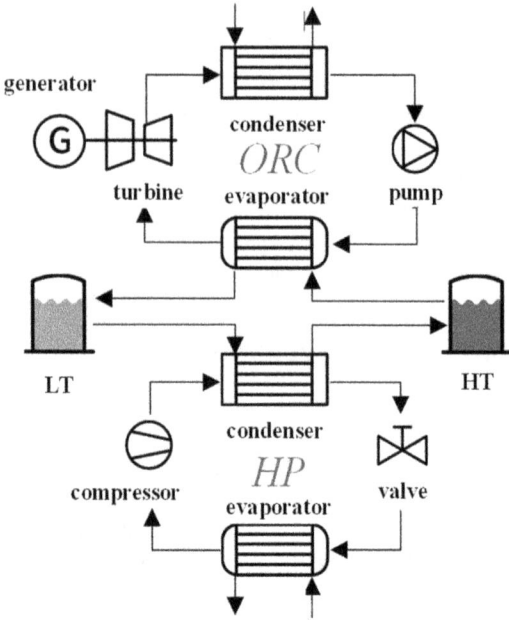

FIGURE 2.1 Typical Rankine CB.

A key architectural consideration involves the potential for component sharing between the charging and discharging cycles. Advanced system designs may utilize reversible machinery that functions as both compressor and expander, significantly reducing capital costs and system complexity. However, such implementations require sophisticated control systems and specialized machinery capable of efficient operation in both modes.

2.2.1.2 Brayton-based system layout

The Brayton CB configuration operates entirely with gaseous working fluids, eliminating the complexity associated with phase change processes while enabling operation at significantly higher temperatures. This configuration typically achieves superior round-trip efficiency due to the higher temperature ratios achievable and the absence of phase change limitations that constrain liquid-vapor systems. The system architecture mirrors that of gas turbine power plants, with the charging cycle implementing a reverse Brayton cycle (gas refrigeration cycle) and the discharging cycle utilizing a conventional Brayton cycle. The absence of phase change processes enables more flexible system design and control, as the thermodynamic states can be varied continuously without concerns about two-phase region limitations. Typical Brayton CB is shown in Figure 2.2.

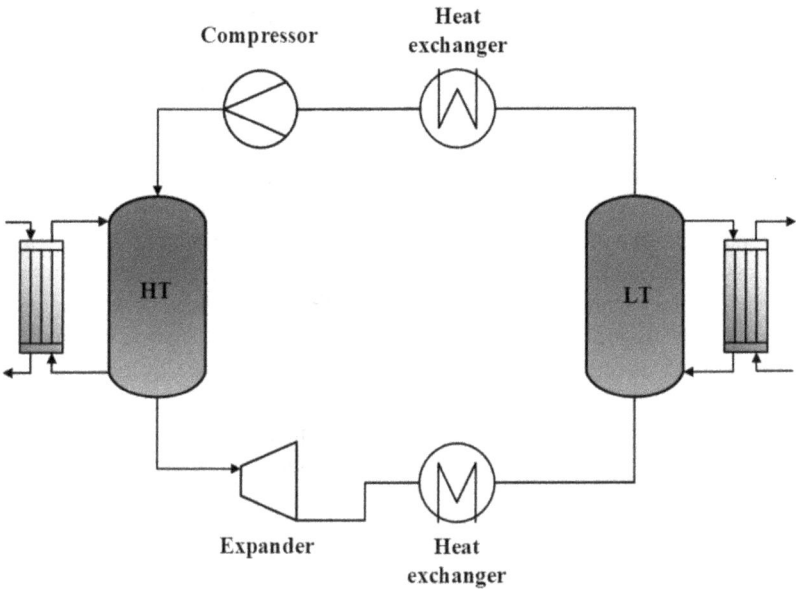

FIGURE 2.2 Typical Brayton CB.

This figure presents the configuration of a typical Brayton-based CB system. Unlike Rankine cycles, the Brayton system operates entirely with gaseous working fluids. The left side shows the reverse Brayton cycle for charging with a compressor, heat exchanger, and expander. The right side illustrates the forward Brayton cycle for discharging. The central thermal storage typically employs packed bed designs using solid materials like rocks or ceramics, enabling operation at temperatures exceeding 500°C for higher efficiency.

The higher operating temperatures, typically exceeding 500°C, necessitate specialized materials and components capable of withstanding elevated thermal stresses. However, these temperature levels enable excellent integration with concentrated solar power systems, industrial HT processes, and other high-grade thermal energy sources. The thermal storage system for Brayton configurations typically employs packed bed designs using solid storage media such as rocks, ceramics, or specialized thermal storage materials. This approach offers excellent thermal stability, relatively low costs, and the ability to operate at HT without degradation concerns.

2.2.1.3 Hybrid and advanced layouts

Advanced system layouts incorporate elements from both Rankine and Brayton configurations to optimize performance for specific applications. These hybrid systems may utilize different thermodynamic cycles for charging and discharging operations, enabling optimization of each process independently. For example, a system might employ a supercritical CO_2 cycle for charging to achieve HT storage while utilizing an ORC for discharging to optimize power generation efficiency at moderate temperatures. Cascaded configurations represent another advanced layout approach, implementing multiple thermodynamic cycles operating at different temperature levels within a single system. This approach enables improved temperature matching with thermal storage systems and can achieve higher overall efficiency by minimizing irreversibilities associated with finite temperature differences. Integration with external thermal sources represents a crucial consideration in advanced system layouts. The incorporation of industrial waste heat, solar thermal energy, or geothermal sources can significantly improve system performance by reducing the temperature lift required during charging operations. Such integration requires careful consideration of thermal interface design, control system complexity, and operational scheduling to maximize benefits while maintaining system reliability.

2.2.2 Storage options

The thermal storage subsystem represents the critical component that determines the energy capacity, storage duration, and overall performance characteristics of CB. Unlike electrochemical storage systems where energy density is primarily determined by battery chemistry, thermal storage systems offer significant flexibility in selecting storage media, configuration, and operating parameters to optimize performance for specific applications.

2.2.2.1 Sensible heat storage systems

Sensible heat storage systems store thermal energy by raising the temperature of a storage medium without undergoing phase change. These systems offer several fundamental advantages, including technological maturity, design simplicity, and operational reliability. The storage capacity is directly proportional to the mass of storage medium, its specific heat capacity, and the temperature rise achieved during charging operations.

Water-based sensible storage systems represent the most mature and widely implemented technology for moderate temperature applications. Pressurized water systems can operate effectively up to approximately 200°C, offering excellent heat transfer characteristics, low cost, and established design methodologies. The high specific heat capacity of water (approximately 4.2 kJ/kg·K) provides favorable volumetric energy density, while the liquid nature enables effective heat transfer through forced or natural convection. For higher temperature applications, alternative sensible storage media become necessary. Thermal oils offer operational capability up to approximately 400°C while maintaining liquid phase, enabling continued utilization of liquid-based heat transfer and storage system designs. However, thermal oils typically exhibit lower specific heat capacities and higher costs compared to water, while also presenting concerns regarding thermal degradation and safety.

Solid media sensible storage systems utilize materials such as rocks, sand, concrete, or specialized ceramics to store thermal energy. These systems can operate at very HT, often exceeding 1000°C, making them particularly suitable for HT CB applications. Packed bed configurations are commonly employed, where the solid storage medium is contained within insulated vessels and heat transfer fluid flows through the packed bed to charge and discharge the storage system. The thermal stratification phenomenon represents a critical consideration in sensible storage system design. During charging and discharging operations, thermal fronts propagate through the storage medium, creating distinct temperature gradients. Proper management of thermal stratification can significantly improve system efficiency by minimizing mixing losses and maintaining HT differentials. However, excessive stratification can lead to non-uniform utilization of storage capacity and reduced effective storage volume.

2.2.2.2 Latent heat storage systems

Latent heat storage systems utilize the phase change enthalpy of storage materials to achieve high energy density storage at relatively constant temperatures. These systems offer several advantages over sensible storage, including higher energy density, more compact system designs, and isothermal charging and discharging characteristics that align well with HP and power cycle requirements.

Phase Change Materials (PCMs) suitable for CB applications must exhibit several critical characteristics: high latent heat of fusion, appropriate melting temperature for the intended application, good thermal stability over many cycles, low cost, and compatibility with containment materials. Organic PCMs such as paraffins offer good thermal properties and stability but are limited to moderate temperatures, typically below 200°C. Inorganic PCMs, including salt hydrates and metallic alloys, can operate at higher temperatures but may present challenges regarding thermal stability, corrosion, and cost.

The selection of appropriate PCMs requires careful consideration of the specific temperature requirements of the CB. For Rankine-based systems operating at moderate temperatures, organic PCMs such as fatty acids, sugar alcohols, or specialized paraffin compounds may be suitable. Higher temperature Brayton systems may require inorganic PCMs such as chloride or fluoride salts, or even metallic PCMs. Heat transfer enhancement represents a critical challenge in latent heat storage systems due to the typically low thermal conductivity of many PCMs. Various enhancement techniques have been developed, including the use of extended surfaces (fins), embedding high-conductivity materials within the PCM, and utilizing encapsulation techniques that increase the surface-to-volume ratio for heat transfer. Encapsulation strategies for PCMs range from macro-encapsulation in containers or tubes to micro-encapsulation where PCM particles are individually encased in protective shells. Macro-encapsulation is simpler and more cost-effective but may suffer from heat transfer limitations. Micro-encapsulation can provide superior heat transfer characteristics and prevent PCM leakage but typically involves higher costs and more complex manufacturing processes.

2.2.2.3 Thermochemical storage systems

Thermochemical storage represents the highest energy density storage option for CB, utilizing reversible chemical reactions to store and release thermal energy. These systems can achieve energy densities significantly exceeding sensible and latent heat storage while potentially offering indefinite storage duration without thermal losses. Common thermochemical storage reactions suitable for CB applications include metal hydride reactions, metal oxide reactions, and salt hydration-dehydration cycles. Metal hydride systems can operate across a wide temperature range and offer excellent reversibility, but typically require expensive materials and present safety concerns regarding hydrogen handling. Metal oxide reactions, such as the cobalt oxide/cobalt system, can operate at HT suitable for Brayton systems but may suffer from sintering and reactivity degradation over cycling. Salt hydration systems, utilizing reactions such as the dehydration of magnesium sulfate or calcium oxide, offer potentially lower costs and good reversibility. However, these systems often require precise water vapor pressure control and may exhibit kinetic limitations that affect charging and discharging rates.

The integration of thermochemical storage with CB requires careful consideration of reaction kinetics, heat and mass transfer limitations, and system complexity. The potential for superior energy density and storage duration must be balanced against increased system complexity, higher costs, and potential reliability concerns.

2.2.3 Working fluid and component selection

The selection of working fluids and system components represents one of the most critical design decisions for CB, directly influencing thermodynamic performance, system complexity, safety characteristics, and economic viability. The working fluid must satisfy the thermal and operational requirements of both charging and discharging cycles while meeting practical constraints regarding availability, cost, safety, and environmental impact [13].

2.2.3.1 Key factors in working fluid selection

The selection of working fluids for CB requires comprehensive evaluation of multiple thermodynamic, environmental, and operational parameters. Critical temperature and pressure represent fundamental constraints that determine the operational envelope of the system. For Rankine-based CB, the working fluid must possess a critical temperature sufficiently higher than the maximum operating temperature to ensure subcritical operation and avoid performance degradation. Conversely, Brayton cycle systems require fluids with appropriate critical pressures that enable efficient HT operation. CO_2 and water demonstrate superior performance in HT applications exceeding 500°C, while organic fluids such as R245fa and R1233zd(E) are optimally suited for moderate temperature ranges between 100°C and 350°C.

Thermal stability constitutes another crucial selection criterion, as working fluids must maintain chemical integrity under prolonged exposure to elevated temperatures without decomposition or property degradation. Organic fluids like R245fa exhibit excellent thermal stability and low toxicity characteristics, making them attractive for medium-temperature applications. However, their operational temperature ceiling of approximately 300°C limits their applicability in HT systems. The viscosity and heat transfer characteristics of working fluids directly influence system efficiency through their impact on pumping power requirements and heat exchanger performance. Lower viscosity fluids generally provide enhanced heat transfer coefficients and reduced friction losses, with R245fa achieving an optimal balance between low viscosity and adequate thermal conductivity.

Environmental considerations have gained paramount importance in working fluid selection, with global warming potential (GWP) and ozone depletion potential (ODP) serving as key evaluation metrics. Environmentally friendly fluids such as R1233zd(E) with low GWP values are increasingly preferred to comply with stringent environmental regulations and sustainability objectives. The economic viability of working fluids, including procurement costs, availability, and compatibility with existing infrastructure, also significantly influences selection decisions for commercial CB implementations.

2.2.3.2 Working fluid categories and applications

Organic working fluids dominate Rankine CB applications due to their favorable thermodynamic properties, safety characteristics, and cost-effectiveness. These fluids typically exhibit low critical pressures, making them suitable for subcritical cycles while providing excellent heat transfer performance at moderate temperatures. R245fa represents one of the most extensively studied organic fluids for ORCs, offering exceptional thermal stability and relatively low toxicity. Its optimal operating temperature range of 100°C to 300°C makes it particularly well-suited for medium-temperature CB systems, though its high boiling point restricts applications in higher temperature regimes.

R1233zd(E) has emerged as a highly attractive alternative due to its low GWP and superior thermal stability characteristics. This fluid demonstrates optimal performance in moderate temperature ranges of 150°C to 350°C and is increasingly utilized in thermally integrated CB. The combination of excellent thermodynamic performance and environmental benefits positions R1233zd(E) as a preferred choice for low-emission applications. Cyclopentane offers economic advantages through its low cost and favorable thermodynamic properties, though its high flammability necessitates additional safety considerations in system design. Despite these limitations, its relatively high energy efficiency makes it viable for low- to medium-temperature applications.

Inorganic working fluids, particularly water and CO_2, are predominantly employed in Brayton cycle-based CB for HT applications. Supercritical CO_2 has gained significant attention in Brayton cycles due to its exceptional performance at elevated temperatures exceeding 500°C. The favorable thermodynamic properties of CO_2, including high density and thermal conductivity, combined with relatively low viscosity, result in reduced pumping losses and enhanced system performance. These characteristics make supercritical CO_2 particularly suitable for high-efficiency heat recovery systems and large-scale energy storage applications.

Water, while serving as a traditional working fluid for numerous heat engine cycles, faces limitations due to its relatively low critical temperature of 374°C, restricting its use in HT applications. Nevertheless, water remains popular in ORCs and lower temperature CB due to its high specific heat capacity, which enables efficient thermal energy storage. The abundant availability, low cost, and well-understood properties of water continue to make it an attractive option for specific applications despite temperature limitations.

2.2.3.3 Advanced fluid optimization strategies

Zeotropic mixtures represent an emerging and promising approach in CB design, combining two or more fluids with different boiling points to enhance heat transfer processes through improved temperature profile matching between the working fluid and heat sources or sinks. This characteristic proves particularly valuable for systems operating across wide temperature ranges, where conventional pure fluids may exhibit suboptimal performance due to temperature mismatches in heat exchangers.

The implementation of zeotropic mixtures can significantly improve heat transfer efficiency by reducing temperature differences and minimizing irreversibilities in heat exchange processes, thereby enhancing overall system performance. Specific mixture combinations such as R1233zd(E)-Cyclopentane and R1234ze(E)-Butane have demonstrated superior performance compared to their pure fluid counterparts in applications involving thermal integration and large temperature spans. These mixtures offer the potential to optimize the thermodynamic cycle by tailoring fluid properties to specific operating conditions and temperature profiles.

The development of advanced fluid screening methodologies represents a critical area for future research, focusing on systematic approaches to identify optimal fluid combinations for specific CB applications. Machine learning algorithms and thermodynamic optimization tools are increasingly being employed to evaluate vast combinations of potential working fluids and mixtures, considering multiple performance criteria simultaneously. Future investigations should emphasize the development of supercritical CO_2 systems and other HT fluids to enhance the performance and scalability of CB across diverse energy storage applications. The integration of fluid property databases with system simulation tools enables comprehensive evaluation of working fluid performance under various operating conditions, facilitating more informed selection decisions. Additionally, the consideration of fluid degradation mechanisms and long-term stability under cycling conditions becomes increasingly important for commercial CB implementations. The development of standardized testing protocols and performance metrics will be essential for establishing industry-wide guidelines for working fluid selection and system optimization.

2.2.4 Component selection

The effectiveness of CB fundamentally depends on the seamless integration of four critical components: turbomachinery for energy conversion, thermal storage systems for heat management, heat exchangers for thermal transfer, and working fluids as energy carriers. The technical selection and compatibility of these components directly determine overall system performance and commercial viability.

2.2.4.1 Turbomachinery technologies

Positive displacement expanders, including scroll, screw, and piston types, operate through periodic volume changes to achieve working fluid compression and expansion. Scroll expanders utilize involute-profile surfaces creating multiple compression chambers, offering compact structure, low noise, and minimal vibration characteristics suitable for small-to-medium power ORC applications. Screw expanders employ meshing rotors for continuous displacement processes, delivering high power density and superior variable-condition adaptability through optimized rotor profiles. However, positive displacement machinery faces limitations, including power output constraints from geometric dimensions, high manufacturing precision requirements, and thermal deformation issues under elevated temperatures.

Turbomachinery proves irreplaceable in large-scale CB. Radial turbines demonstrate distinct advantages under medium-flow, high-pressure-ratio conditions due to their relatively simple impeller structure and reduced sensitivity to inlet conditions, making them preferred for supercritical CO_2 cycles. Axial turbines achieve higher efficiency in high-flow applications but face increased complexity and efficiency degradation under variable operating conditions. Current technical challenges encompass HT material limitations, droplet impact erosion under two-phase flow conditions, and the need for sophisticated control mechanisms. Innovative developments include liquid injection technology for temperature control, ceramic-based composite materials enabling higher operating temperatures, and reversible turbomachinery designs reducing system investment costs.

2.2.4.2 Thermal storage systems

Solid-state thermal storage systems achieve heat storage through sensible heat changes in solid materials, offering technological maturity, cost-effectiveness, and high reliability. Natural rock materials, including basalt and granite, provide excellent thermal stability and economic advantages, though relatively low storage density limits system compactness. Synthetic materials such as alumina ceramic spheres offer superior performance but at higher costs. Packed bed structure design critically influences system performance, requiring particle size selection that balances heat transfer effectiveness against pressure drop losses through strategic multi-stage configurations.

PCMs utilize solid-liquid phase transitions for high-density latent heat storage, offering constant temperature operation advantages. Inorganic salt PCMs, including sodium nitrate and potassium nitrate, excel in medium-temperature ranges with high latent heat values and good chemical stability, though supercooling phenomena and corrosiveness present application challenges. Organic PCMs provide good chemical stability but suffer from low thermal conductivity and flammability issues. Composite thermal storage technology combines different mechanisms to leverage individual advantages, with form-stable PCM composites preventing liquid leakage while providing enhanced heat transfer areas, though manufacturing complexity and long-term stability verification remain challenging.

2.2.4.3 Heat exchanger technologies

Compact heat exchangers occupy critical positions through high specific surface areas and excellent heat transfer performance. Plate-fin heat exchangers enhance heat transfer through corrugated fins, achieving extremely high volumetric heat transfer coefficients particularly suitable for gas-gas applications, though complex flow channels can create distribution irregularities. Plate heat exchangers utilize stacked thin-plate structures offering high efficiency, compact design, and convenient maintenance, with excellent sealing performance making them ideal for organic working fluid applications, though HT sealing and plate deformation require attention.

Advanced heat exchange technologies represent cutting-edge development directions. Printed Circuit Heat Exchangers manufactured through chemical etching or diffusion bonding achieve extremely high heat transfer coefficients and compactness, demonstrating unparalleled advantages under extreme conditions like supercritical CO_2, though complex manufacturing processes and high costs present limitations. Microchannel technology intensifies heat transfer through reduced dimensions but requires extremely high working fluid cleanliness and strict manufacturing tolerances. Material science applications, including silicon carbide ceramics and metal matrix composites, enable performance enhancements for extreme conditions, though ceramic brittleness, complex manufacturing processes, and interfacial bonding strength represent ongoing technical challenges.

2.2.4.4 Working fluid selection influence

Working fluids as energy carriers directly influence heat and mass transfer processes and energy conversion efficiency. Critical temperature and pressure determine application ranges and system pressure levels, while molecular weight affects density and flow characteristics influencing equipment sizing. In Brayton cycle applications, helium provides minimal molecular weight and excellent heat transfer performance for high-performance systems but faces scarcity and cost limitations, while argon offers abundant resources and lower costs for practical applications. Rankine cycle selection proves more complex, with water vapor remaining primary for HT applications despite high saturation pressure and corrosiveness requirements.

Organic working fluid development addresses medium-low temperature applications, with traditional refrigerants facing phase-out pressure due to environmental concerns. Next-generation HFO compounds feature zero ODP and extremely low GWP as replacement solutions, though thermal stability and material compatibility require investigation. Natural working fluids, including CO_2, ammonia, and hydrocarbons, have received renewed attention despite safety challenges from flammability or toxicity. Working fluid mixtures achieve enhanced performance through component synergistic interactions, with non-azeotropic mixtures reducing heat exchange irreversible losses and azeotropic mixtures maintaining operational simplicity, though complex calculations and composition control present engineering challenges requiring comprehensive theoretical research and validation.

CB success depends on sophisticated integration of these four core components, with each technology's advancement and their synergistic optimization determining overall performance and commercial viability for large-scale energy storage applications.

2.3 OPTIMIZATION APPROACHES

Current research on CB predominantly treats them as independent system modules, focusing on optimizing their overall configuration and performance, such as improving efficiency and economic viability through advancements in thermodynamic cycles, thermal storage materials, equipment, or system integration. However, few studies have deeply explored the sensitivity of internal parameters and their impact on system performance, such as heat source temperature, cold source temperature, and thermal storage temperature.

2.3.1 Case study

2.3.1.1 System introduction

This chapter proposes a Rankine CB, with R1233zd(E) as the working fluid for both the HP and ORC. This study focuses on the resource utilization pathway for atmospheric-pressure wastewater waste heat in the temperature range of 60–85°C within industrial integrated energy system, with the specific structure illustrated in Figure 2.3.

This figure shows the detailed system layout for the case study Rankine CB. The HP cycle on the left absorbs heat from industrial wastewater at 60–85°C and delivers it to the thermal storage system through the condenser. The ORC on the right extracts heat from the HT tank to generate electricity. The system uses R1233zd(E) as the working fluid for both cycles, achieving efficient recovery and storage of low-grade industrial waste heat.

In HP, the low-pressure refrigerant first absorbs heat isobarically in the evaporator before entering the compressor, where it undergoes non-isentropic

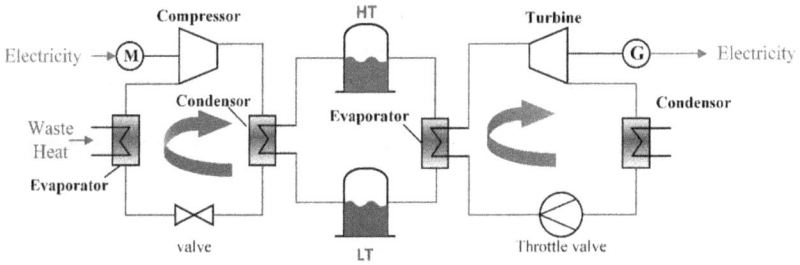

FIGURE 2.3   System layout of the case-study Rankine CB.

compression, increasing both pressure and temperature. It then flows into the condenser, releasing heat to the thermal storage medium. The HT storage medium is stored in the HT (HT storage tank), while the LT medium is stored in the LT (LT storage tank). In ORC, the HT medium in the HT releases latent heat to the evaporator, becoming a LT medium and returning to the LT for storage. The organic working fluid absorbs heat isobarically in the evaporator, vaporizing, and then enters the turbine to expand and perform work. The exhaust steam after work condenses isobarically in the condenser, while the ambient cold source absorbs the condensation heat in the condenser. The specific process is illustrated in Figure 2.4.

This temperature-entropy diagram illustrates the thermodynamic cycles of the Rankine CB. The HP cycle operates counterclockwise, with the working fluid absorbing heat in the evaporator, being compressed, releasing heat in the

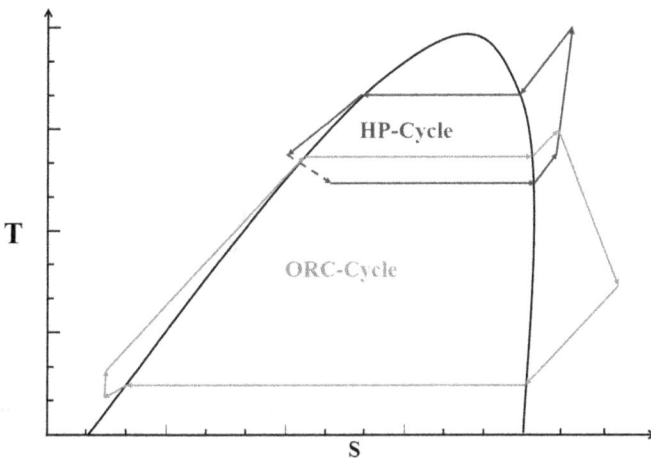

FIGURE 2.4   Thermodynamic cycle temperature-entropy diagram of Rankine CB.

condenser, and throttling back to the initial state. The ORC runs clockwise, with the working fluid being pressurized by the pump, evaporated in the evaporator, expanded in the turbine, and condensed to complete the cycle.

2.3.1.2 Carnot battery thermodynamic state solution

To solve the design condition of CB, it is first necessary to preliminarily set boundary conditions and the thermodynamic properties of the equipment to establish a comprehensive thermodynamic system model. In the model, each piece of equipment is interconnected through its respective thermodynamic equations, forming a complex thermodynamic network. Within this network, each node must satisfy its own constraints while also adhering to the conservation of mass, energy, and momentum. The solution of the network is based on the Newton-Raphson algorithm, which iteratively optimizes to gradually approach the thermodynamic equilibrium state of the system. In each iteration, the system calculates the residuals of the thermodynamic variables based on the current state and updates the Jacobian matrix and the current residual values, as shown in the following equation:

$$J^{(k)}\Delta X^{(k)} = -F\left(X^{(k)}\right) \tag{2.17}$$

$$X^{(k+1)} = X^{(k)} + \alpha\Delta X^{(k)}$$

Until the preset convergence condition is met, the solution is complete. The convergence condition is as follows:

$$\left|F\right|_2 < 10^{-6} \tag{2.18}$$

$$\max\left(\frac{\left(\Delta X_i\right)}{\left|X_i\right| + \varepsilon}\right) < 10^{-4} \tag{2.19}$$

To solve the off-design condition of CB, by setting core parameters such as compressor isentropic efficiency, heat exchanger heat transfer coefficient, and working fluid mass flow rate under design conditions, the thermal states (temperature, pressure, enthalpy, etc.) of each system node can be solved. Meanwhile, the equipment geometric parameters and characteristic curve clusters obtained from the solution serve as reference data for off-design condition solutions.

In the off-design condition solution stage, the geometric dimensions of heat exchangers and the inherent characteristics of equipment are fixed and frozen. Equipment characteristic curves such as heat exchanger characteristic heat transfer coefficient curves, compressor efficiency curves, and expander efficiency curves are introduced. Interpolation techniques are utilized to obtain the nonlinear characteristics of heat exchangers and turbomachinery, which can provide support for thermal network off-design condition solutions. When the operating points of CB components deviate from the design conditions, parameters such as pressure drop and isentropic efficiency set during design condition solution are no longer fixed. Equipment characteristic curves are introduced, and the Newton-Raphson method is used for iterative solution according to working fluid mass flow rate variations to obtain the actual performance of equipment and CB thermodynamic cycle parameters under off-design conditions.

2.3.1.3 Variable setting

When designing the key parameters of CB, it is necessary to consider the interaction between boundary conditions and equipment performance. These parameters are tightly coupled and subject to complex constraints. Therefore, variable transformations are applied to most design variables. The specific key design parameters are shown in Table 2.1.

Uniform sampling was performed for eight variables within their respective specific intervals. The initial sample space exhibited non-uniform characteristics, so non-uniform parameter samples were discarded. After sample compression, the upper and lower bounds of the parameters were redefined, forming a sample space of 8000 samples that better aligns with the thermodynamic characteristics of the CB The parameters from the resampled sample space were input into CB model for solution, to investigate the impact of key design parameters on the characteristics of the CB under design conditions:

TABLE 2.1   Specific key design parameters in CB

SYMBOL	TRANSFORMATION RELATIONSHIP	CONSTRAINT
ΔT^{hs}	$\Delta T^{hs} = T_{in}^{hs} - T_{out}^{hs}$	[10, 25]
ΔT_1^{hp}	$\Delta T_1^{hp} = T_{out}^{hs} - T_{evap}^{hp}$	[5, 10]
ΔT_2^{hp}	$\Delta T_2^{hp} = T_{cond}^{hp} - T_{evap}^{hp}$	[40, 60]
ΔT_3^{hp}	$\Delta T_3^{hp} = T_{cond}^{hp} - T_h^{sto}$	[5, 10]
ΔT^{sto}	$\Delta T^{sto} = T_h^{sto} - T_c^{sto}$	[10, 25]
ΔT_1^{orc}	$\Delta T_1^{orc} = T_h^{sto} - T_{evap}^{orc}$	[5, 10]
ΔT_2^{orc}	$\Delta T_2^{orc} = T_{cond}^{orc} - T_{out}^{cs}$	[5, 10]
T_{hs}		[60, 85]

2.3.1.4 Parameter sensitivity analysis

The output values from the aforementioned CB multi-condition performance library are subjected to inner product calculations with the Fourier basis functions in the sampling method to obtain the corresponding Fourier coefficients, as shown in the following equation:

$$c_\omega = \frac{1}{\pi} \int_0^{2\pi} f(x(s)) e^{-i\omega s} ds, \quad \omega \in N^* \tag{2.20}$$

where $***\omega$ is the frequency, $x(s)$ is the input sample, and $f(x(s))$ is the model output.

For each input variable, the calculation formulas for first-order sensitivity index and total sensitivity index are as follows:

First-order sensitivity index:

$$S_{x_i}^{FAST} = \frac{\sum_{l=1}^\infty |c_{\omega_i}|^2}{\sum_{\omega=1}^\infty |c_\omega|^2} \tag{2.21}$$

Total sensitivity index:

$$ST_{x_i}^{FAST} = \frac{\sum_{\omega \geq \omega_i} |c_\omega|^2}{\sum_{\omega=1}^\infty |c_\omega|^2} \tag{2.22}$$

where c_{ω_i} is the Fourier coefficient corresponding to input variable x_i, and ω_i is the corresponding frequency. By comparing the sensitivity indicators of various design parameters on performance metrics, the key design parameters among CB design parameters that have significant impact on performance can be identified. After finishing the parameter sensitivity analysis, we found that heat source temperature, heat source temperature difference, and HP temperature rise are key variables.

2.3.1.5 Influence of heat source temperature

The effect of heat source temperature on the key performance indicators of CB is shown in Figure 2.5. As the heat source temperature increases, CB's performance metrics exhibit significant improvements: η_{orc} mean efficiency rises from 6.8% to 10.1%, η_{rt} from 34.3% to 67.9%, and η_{ex} from 25.3% to

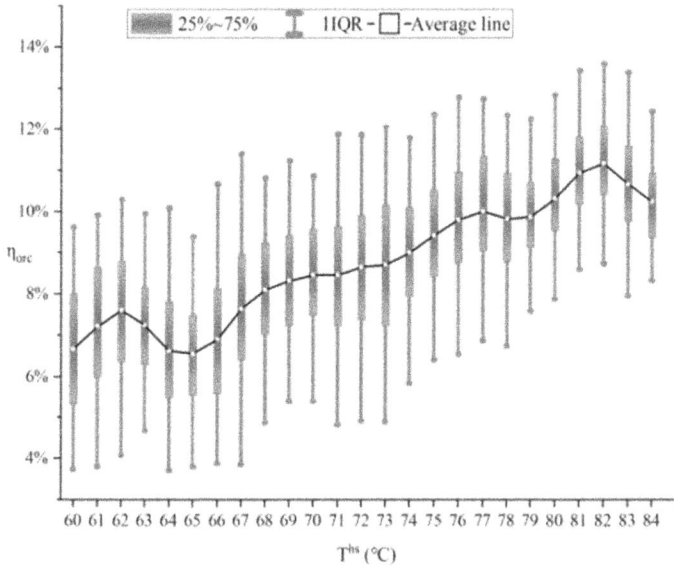

(a)

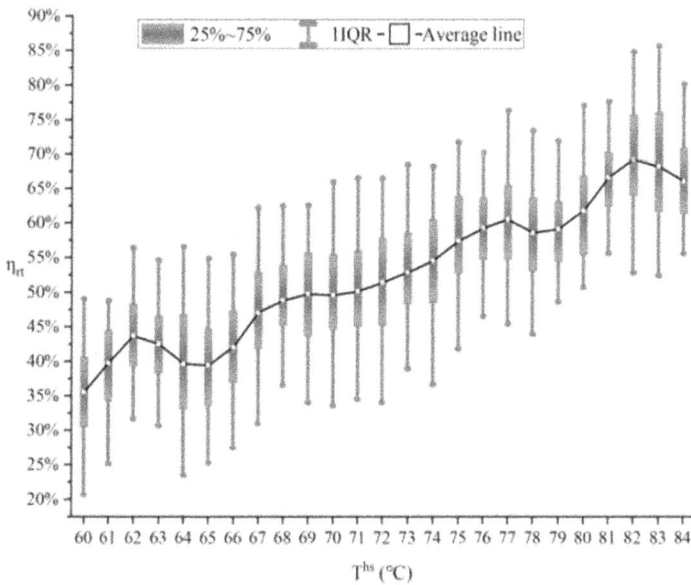

(b)

FIGURE 2.5 Influence of heat source temperature on the parameters of CB. (a) ORC Efficiency. This subplot displays the variation of the Organic Rankine Cycle

(Continued)

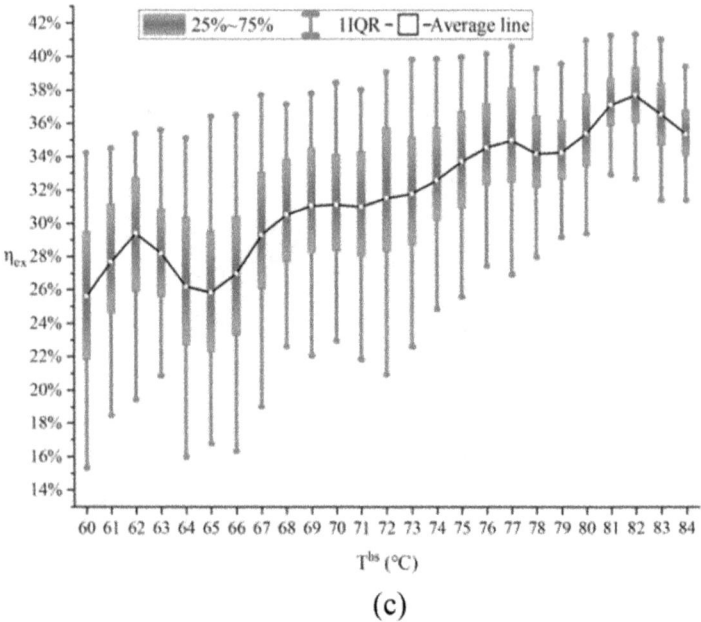

(c)

FIGURE 2.5 (Continued)

(ORC) efficiency. As the heat source temperature(*Ths*) increases from 60°C to 84°C, the mean efficiency of the ORC exhibits a steady upward trend, rising from approximately 6.8% to 10.1%. This improvement is thermodynamically attributed to the enhanced thermal quality of the heat source; a higher *Ths* allows the HP to deliver higher-grade heat, which subsequently improves the operating conditions and efficiency of the ORC subsystem. The box plots indicate relatively narrow interquartile ranges (IQR), suggesting a consistent positive correlation across the parameter space. (b) Round-trip Efficiency. This subplot shows the impact of *Ths* on the system's round-trip efficiency. A significant monotonic increase is observed, with the average round-trip efficiency improving dramatically from 34.3% to 67.9%. The wider spread of the whiskers and boxes at higher temperatures indicates that while higher heat source temperatures generally favor round-trip efficiency, the system becomes more sensitive to other design variables (such as component efficiencies) in the high-temperature region. (c) Exergy Efficiency. The exergy efficiency, depicted in this subplot, follows a similar increasing trend, growing from 25.3% to 35.4%. This metric confirms that raising the heat source temperature not only improves energy conversion ratios but also enhances the thermodynamic quality of the energy preservation, reducing irreversible losses during the heat recovery and storage process.

35.4%. Additionally, the round-trip efficiency generally increases with higher heat source temperatures. This is attributed to the independent operation of HP and ORC. Elevated heat source temperatures enhance the thermal quality available to HP, consequently improving the performance of the ORC. This is because in the CB we built, the HP and ORC operate independently of each other, and the increase in heat source temperature enhances the grade of heat available for the HP cycle.

These three subfigures demonstrate how heat source temperature affects key performance parameters. The ORC efficiency increases from approximately 7% to 10%, round-trip efficiency rises from 34% to 68%, and exergy efficiency improves from 25% to 35% as heat source temperature increases. All performance indicators show monotonic improvement with higher heat source temperatures due to enhanced thermal energy quality available for the cycles.

2.3.1.6 Influence of heat source temperature difference

The impact of the heat source temperature difference on the key performance indicators of a thermally integrated CB is depicted in Figures 2.6. As the heat source temperature difference increases, the performance metrics of the CB exhibit a declining trend: η_{rt} mean efficiency decreases from 62.6% to 45.8%, η_{orc} from 10.4% to 7.5%, and η_{ex} from 36.5% to 27.8%. This reduction is attributed to the operational characteristics of the CB, where a larger heat source temperature difference requires greater work input in HP to transfer heat, consequently leading to a decline in overall efficiency. This is because in a CB, the higher the temperature difference of the heat source, the more work is required in the HP cycle to transfer heat. Therefore, it will show a downward trend.

These three subfigures show the impact of heat source temperature difference on system performance. Round-trip efficiency decreases from 63% to 46%, ORC efficiency drops from 10.4% to 7.5%, and exergy efficiency declines from 36.5% to 27.8% as temperature difference increases. The declining trend occurs because larger temperature differences require more compression work in the HP cycle, reducing overall system efficiency.

2.3.1.7 Influence of HP temperature rise

The impact of HP temperature difference on the key performance indicators of CB is shown in Figures 2.7. As HP temperature rise increases, the COP decreases from 7.6 to 4.8, while η_{orc} mean efficiency rises from 7.0% to 10.2%. The COP generally decreases with an increase in the heat source temperature difference. This is because, in this CB, the thermal storage temperature is primarily determined by constraints. To maintain a stable thermal storage

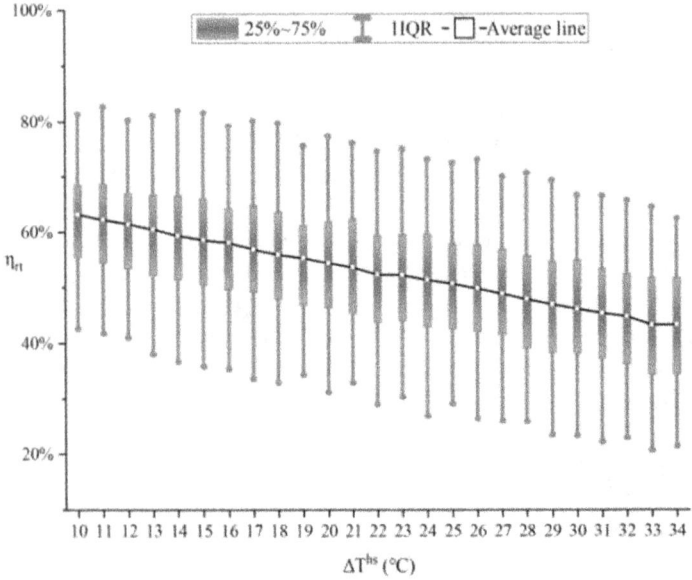

(a)

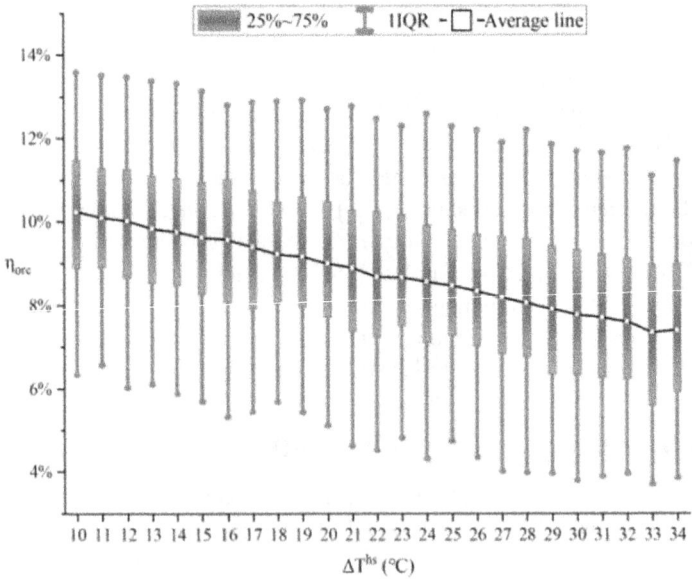

(b)

FIGURE 2.6   Influence of heat source temperature difference on the parameters of CB. (a) Round-trip Efficiency. This subplot illustrates a clear declining trend in

(Continued)

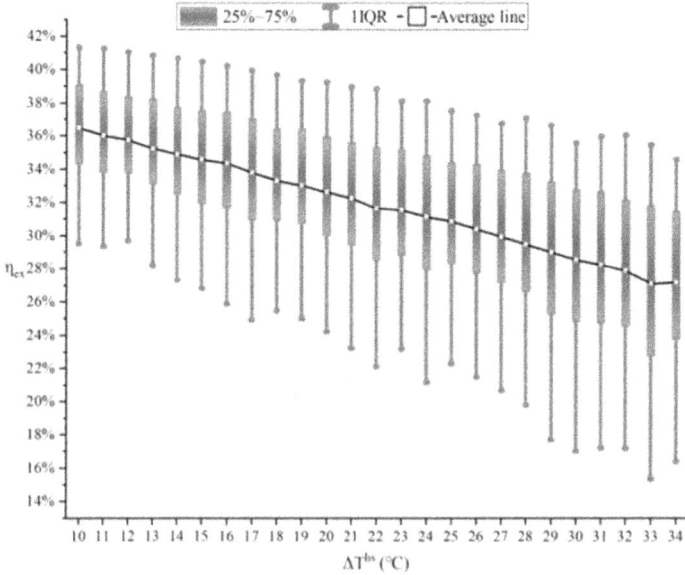

(c)

FIGURE 2.6 (Continued)
round-trip efficiency as the temperature difference increases. The mean round-trip efficiency decreases from 62.6% to 45.8% across the sampled range. This performance degradation occurs because a larger ΔThs necessitates a greater pressure lift in the compressor. The HP requires significantly more work input to transfer heat across a larger temperature gradient, which lowers the overall energy output-to-input ratio of the battery. (b) ORC Efficiency. The ORC efficiency also demonstrates a downward trend, dropping from 10.4% to 7.5% as ΔThs increases. While the HP and ORC cycles operate somewhat independently, the constraints imposed by a larger temperature glide at the heat source boundary indirectly affect the optimal operating points of the cycle, leading to reduced thermal conversion efficiency in the discharge phase. (c) Exergy Efficiency. Consistent with the other indicators, the exergy efficiency decreases from 36.5% to 27.8%. The negative slope in the box plots highlights the thermodynamic penalty associated with large temperature differences. The increase in irreversible losses (exergy destruction) during the heat transfer process at larger ΔThs values is the primary driver for this decline. The relatively uniform size of the boxes suggests this negative correlation is robust across various system configurations.

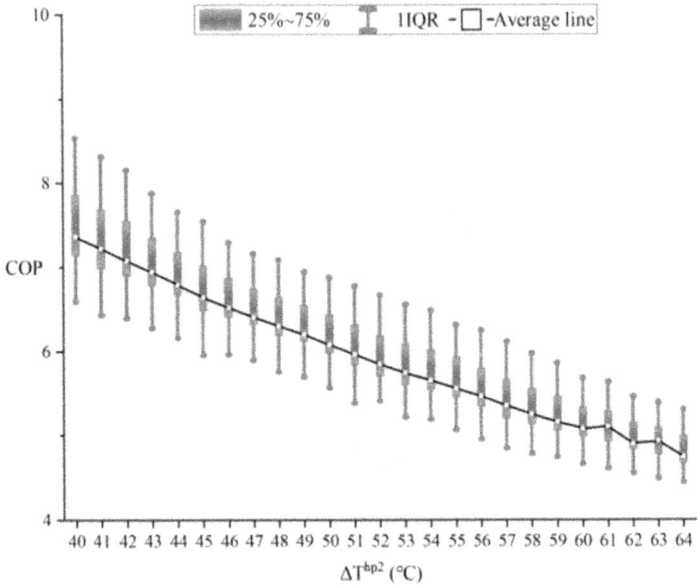

(a)

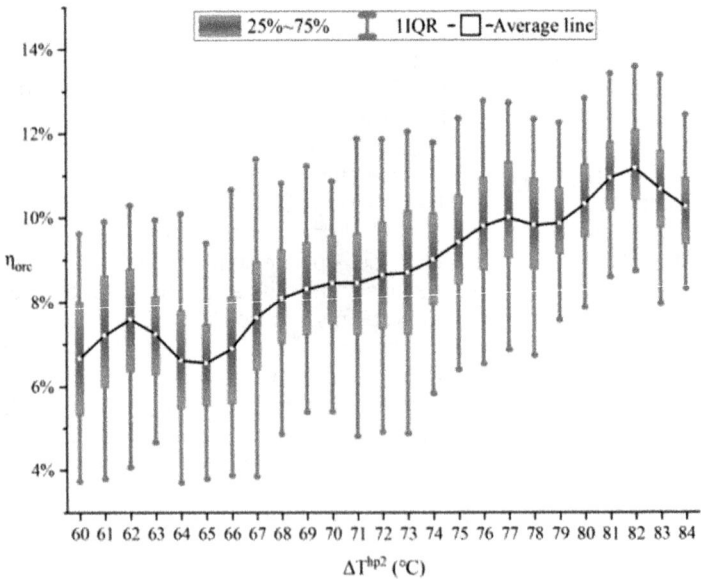

(b)

FIGURE 2.7 Influence of HP temperature rise on the parameters of CB.

temperature, the condensation temperature must also be maintained at a certain high level. An increase in HP temperature rise implies a reduction in the evaporation temperature, which increases the compressor work input, consequently leading to a decrease in COP. Overall, the COP decreases as the temperature difference of the heat source increases. This is because in this CB, the heat storage temperature is basically determined by constraints; to ensure the stability of the heat storage temperature, the condensation temperature must also be maintained at a certain high level. An increase in the temperature rise of the HP means a decrease in the evaporation temperature. In order to maintain a stable heat exchange rate of the evaporator, it is necessary to reduce the mass flow rate of the working fluid cycle. The difference between the condensation pressure and the evaporation pressure increases, which leads to an increase in the power consumption of the compressor and thus a decrease in the COP.

These two subfigures illustrate how HP temperature rise affects performance. The coefficient of performance decreases from 7.6 to 4.8, while ORC efficiency increases from 7% to 10.2%. The opposite trends result from system thermodynamic constraints: higher temperature rise reduces evaporation temperature, increasing compressor work and lowering COP, but simultaneously provides higher-grade heat for the ORC, improving its efficiency.

2.3.1.8 Off-design analysis

When fluctuations occur in the operation of actual industrial integrated energy systems, causing changes in the boundary conditions of the CB, the impacts induced by boundary condition disturbances can be balanced by adjusting the working fluids of the HP and ORC in the energy storage cycle and energy discharge cycle of the CB. In addition, variations in the working fluid flow rates will lead to performance changes of components such as evaporators, condensers, compressors, and expanders in the CB, thereby affecting the performance of the CB. Taking a group of CB in the aforementioned feasible parameter space as an example, when the mass flow rates of the HP and ORC fluctuate within ±30% of their design conditions, the changes in system performance are as follows.

Figures 2.8 and 2.9 reflect the variations of COP and η_{orc} under off-design conditions. Since the HP and ORC operate relatively independently, the performance of the HP does not directly induce changes in the performance of the ORC. Instead, it indirectly affects the performance of the ORC through the heat transferred to the heat storage system and changes in the mass flow rate of the working fluid in the heat storage system, and the same applies to the impact of the ORC on the HP.

This three-dimensional surface plot shows how the coefficient of performance varies with mass flow rate changes in both the HP and ORC cycles

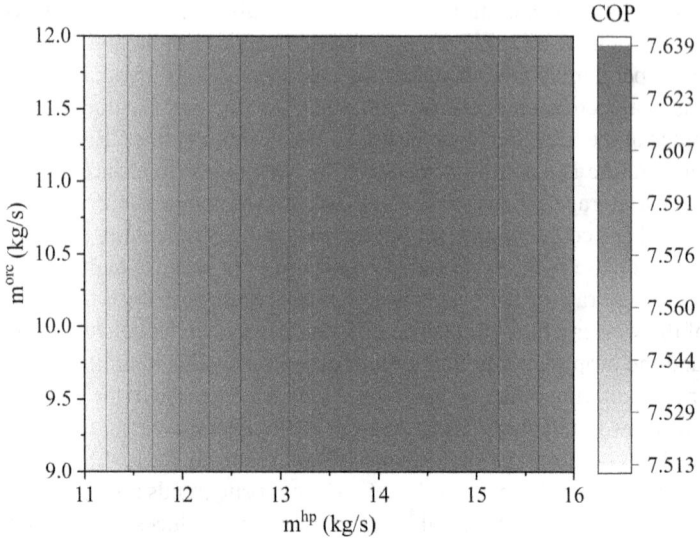

FIGURE 2.8   Variation of COP under off-design conditions.

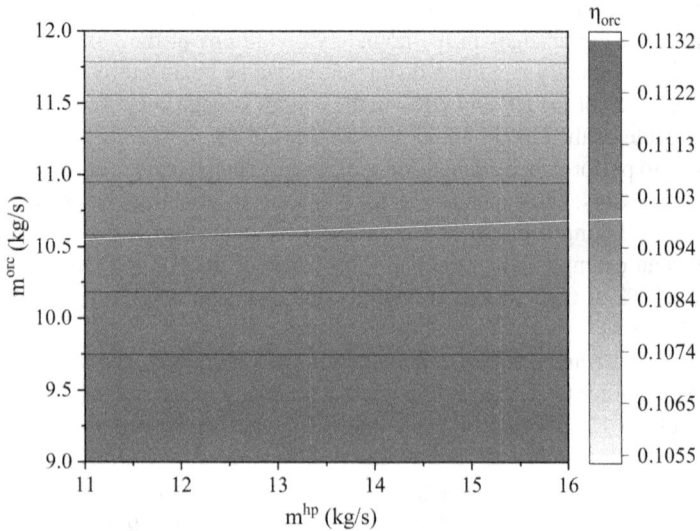

FIGURE 2.9   Variation of η_{orc} under off-design conditions.

within ±30% of design values. The surface demonstrates that COP is primarily influenced by the HP's own mass flow rate, showing relatively weak coupling with the ORC cycle. The performance decreases as HP mass flow rate increases from design conditions.

This three-dimensional surface plot illustrates ORC efficiency variations under off-design conditions as both cycle mass flow rates fluctuate within ±30% of design values. The surface shows that ORC efficiency is mainly affected by its own mass flow rate, exhibiting an optimal operating point. The HP mass flow rate has a more indirect influence, primarily affecting ORC performance through changes in thermal storage heat transfer.

Figures 2.10 and 2.11 reflect the variations of η_{rt} and η_{ex} under off-design conditions. It can be seen that the round-trip efficiency and exergy efficiency are jointly affected by the mass flow rates of the HP and ORC. Among them, the impact of ORC mass flow rate fluctuations on the round-trip efficiency and exergy efficiency is more significant than that caused by the HP mass flow rate fluctuations. The main reason is that the characteristic curve of the expander used in the ORC is more significantly affected by changes in the working fluid mass flow rate compared to the characteristic curve of the compressor in the HP.

This three-dimensional surface plot displays round-trip efficiency variations under off-design conditions. The surface reveals that round-trip efficiency is jointly influenced by both cycle mass flow rates, with ORC mass flow rate

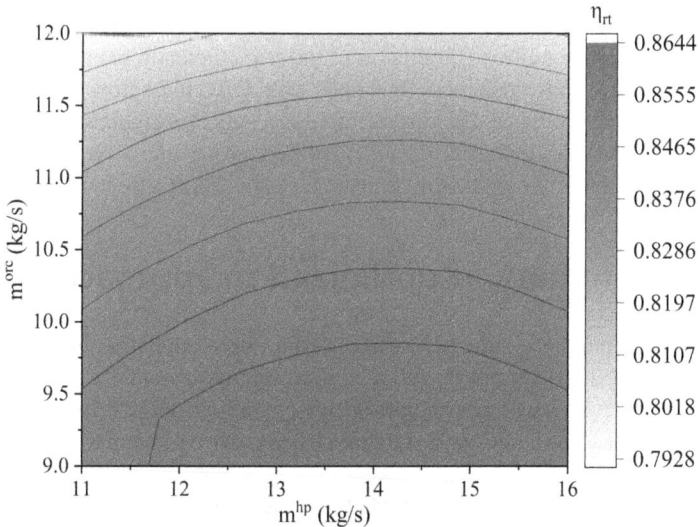

FIGURE 2.10 Variation of η_{rt} under off-design conditions.

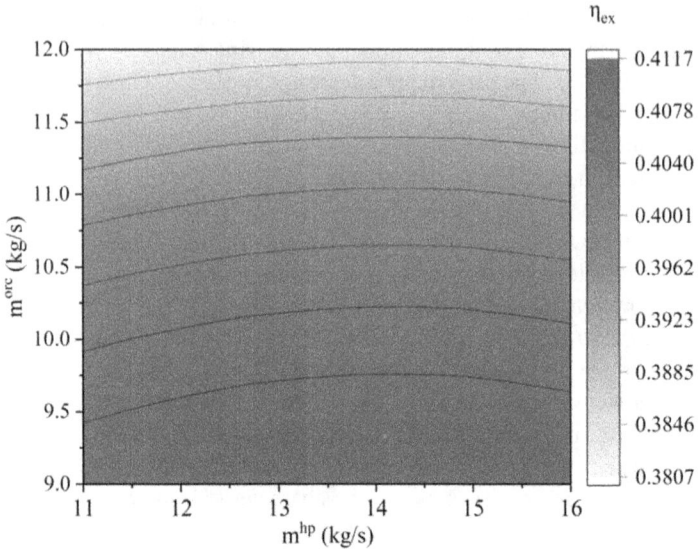

FIGURE 2.11   Variation of η_{ex} under off-design conditions.

having a more significant impact. This reflects the expander's characteristic curve being more sensitive to flow rate changes compared to the compressor, resulting in larger efficiency variations in the ORC direction.

This three-dimensional surface plot depicts exergy efficiency variations under off-design conditions. Similar to round-trip efficiency, exergy efficiency is influenced by both cycle mass flow rates, with ORC mass flow rate showing more pronounced effects. The surface topology indicates significant increases in irreversible losses under off-design conditions, particularly requiring attention to ORC operation regulation to maintain high exergy efficiency.

2.3.2 Summary of optimization approaches

This chapter mainly conducts research on the thermodynamic characteristic analysis and parameter optimization of CB, aiming to obtain a performance database of CB under design and off-design conditions under different boundary conditions. The main work of this chapter can be summarized into the following three parts:

First, based on the electricity-heat-electricity energy conversion path in CB, the energy efficiency and exergy efficiency indicators of CB with general configurations are analyzed. Furthermore, for heat-integrated CB, considering equipment performance changes under off-design conditions and differences in energy storage and discharge time, dynamic energy efficiency and exergy efficiency indicators of CB are defined.

Second, a characteristic analysis method for CB based on the construction of a multi-condition performance database is proposed. The parameters of each link (energy storage, heat storage, and energy discharge) in CB are divided into boundary condition parameters (such as waste heat resource temperature), equipment performance parameters (such as compressor isentropic efficiency), and design parameters (such as HP evaporation temperature). The design parameters like HP evaporation temperature are converted into temperature differences to embed thermodynamic constraints. A parameter space is constructed through joint sampling of variables, and parameter space compression and resampling are performed based on thermodynamic constraints and model convergence to obtain a multi-condition performance database of CB, supporting the characteristic analysis of CB.

Finally, the multi-condition characteristics of CB for recovering wastewater waste heat resources at 60–85°C are analyzed. The key design variables affecting the performance of CB are identified as heat source temperature, heat source temperature difference, and HP temperature rise, and the characteristics of CB under design and off-design conditions are analyzed. The analysis results show that under design conditions, the heat source temperature is positively correlated with ORC efficiency, round-trip efficiency, and exergy efficiency; the heat source temperature difference is positively correlated with ORC efficiency, round-trip efficiency, and exergy efficiency; the HP temperature rise is negatively correlated with COP and positively correlated with ORC efficiency. Under off-design conditions, COP and ORC efficiency are mainly affected by the working fluid mass flow rate in their respective cycles, while round-trip efficiency and exergy efficiency are jointly affected by the working fluid mass flow rates of the HP and ORC, with the impact of ORC working fluid mass flow rate being more significant.

REFERENCES

[1] Gibb, B. C. (2025). Carnot batteries for dispatchable renewables. *Nature Chemistry*, *17*(5), 629–631.
[2] McTigue, J. D., White, A. J., & Markides, C. N. (2015). Parametric studies and optimisation of pumped thermal electricity storage. *Applied Energy*, *137*, 800–811.

[3] Poletto, C., Dumont, O., De Pascale, A., Lemort, V., Ottaviano, S., & Thomé, O. (2024). Control strategy and performance of a small-size thermally integrated Carnot battery based on a Rankine cycle and combined with district heating. *Energy Conversion and Management, 302*, 118111.

[4] Thess, A. (2013). Thermodynamic efficiency of pumped heat electricity storage. *Physical Review Letters, 111*(11), 110602.

[5] White, A., Parks, G., & Markides, C. N. (2013). Thermodynamic analysis of pumped thermal electricity storage. *Applied Thermal Engineering, 53*(2), 291–298.

[6] White, A., McTigue, J., & Markides, C. (2014). Wave propagation and thermo-dynamic losses in packed-bed thermal reservoirs for energy storage. *Applied Energy, 130*, 648–657.

[7] Yang, H., Li, J., Ge, Z., Yang, L., & Du, X. (2022). Dynamic characteristics and control strategy of pumped thermal electricity storage with reversible Brayton cycle. *Renewable Energy, 198*, 1341–1353.

[8] Pecchini, M., Peccolo, S., Benato, A., De Vanna, F., & Stoppato, A. (2024). Analysis of the discharge process of a TES-based electricity storage system. *Journal of Energy Storage, 100*, 113518.

[9] Sun, R., Zhao, Y., Liu, M., & Yan, J. (2022). Thermodynamic design and optimi-zation of pumped thermal electricity storage systems using supercritical carbon dioxide as the working fluid. *Energy Conversion and Management, 271*, 116322.

[10] Li, W., Zhang, L., Deng, Y., & Zeng, M. (2024). Thermo-economic assessment of a salt hydrate thermochemical energy storage-based Rankine Carnot battery system. *Energy Conversion and Management, 312*, 118564.

[11] Iqbal, Q., Fang, S., Xu, Z., Yao, Y., Song, J., Qiu, L., … & Wang, K. (2024). Techno-economic comparison of high-temperature and sub-ambient tempera-ture pumped-thermal electricity storage systems integrated with external heat sources. *Journal of Energy Storage, 89*, 111630.

[12] Shamsi, S. S. M., Barberis, S., Maccarini, S., & Traverso, A. (2024). Thermo-economic performance evaluation of thermally integrated Carnot battery (TI-PTES) for freely available heat sources. *Journal of Energy Storage, 97*, 112979.

[13] Wu, D., Ma, B., Zhang, J., Chen, Y., Shen, F., Chen, X., … & Yang, Y. (2024). Working fluid pair selection of thermally integrated pumped thermal electricity storage system for waste heat recovery and energy storage. *Applied Energy, 371*, 123693.

Modeling of Integrated Energy Systems

3

Yihui Mao, Wenxuan Guo, Xueru Lin, Xiaojie Lin, and Wei Zhong

3.1 OVERVIEW AND MODELING OF INTEGRATED ENERGY SYSTEMS (IES)

3.1.1 Overview of IES modeling

Energy serves as the crucial material basis and a key engine for global economic and social advancement. However, since the 19th century, the unchecked extraction and consumption of fossil fuels have accelerated global climate change, presenting a serious threat to human survival and other life forms. This environmental crisis has led to a worldwide consensus on the urgent need to achieve low-carbon energy goals and accelerate the energy transition.

In this context, China has demonstrated a firm commitment to a low-carbon development path. In September 2020, the nation declared its ambitions to "peak carbon dioxide emissions before 2030 and strive to achieve carbon neutrality before 2060." Furthermore, the report from the 20th National Congress of the Communist Party of China highlighted the mandate to "deepen

DOI: 10.1201/9781003630821-3

the energy revolution, strengthen the clean and efficient use of coal, and accelerate the planning and construction of a new energy system."

Traditional energy systems have historically relied on coal-fired power generation, distributing electricity and heat to consumers via extensive power grids and heating networks. While these architectures have ensured reliable energy security for societal needs, their inherent reliance on fossil fuels has resulted in severe environmental degradation and introduced several critical operational challenges:

1. Integration of Renewable Energy: Difficulty in accommodating the intermittent nature of renewable energy sources;
2. System Flexibility and Adaptability: Insufficient ability to rapidly adjust to fluctuating supply and demand conditions;
3. Market Access: Restrictions hindering the participation of new energy technologies and market players;
4. Energy Utilization Efficiency: Overall low efficiency in the conversion and end-use of energy.

Given these limitations, a shift away from coal-power-dominated traditional energy systems is essential.

The IES, defined by its core principles of multi-energy complementarity and coordinated optimization, is widely recognized as a pivotal strategy for decarbonizing the energy sector and substantially improving energy efficiency. An IES utilizes advanced technology and innovative management paradigms to integrate diverse energy carriers. This approach facilitates synergistic management, optimized operation, and coordinated planning across heterogeneous energy subsystems.

Fundamentally, an IES is structured around the power system, incorporating multiple other energy subsystems. All system elements are seamlessly coordinated and optimized throughout the planning, construction, and operation phases, thereby forming a unified system for energy production, supply, and consumption.

Developing a robust mathematical model that accurately captures the physical behavior of an IES is the prerequisite for all subsequent operational and optimization decisions. For modeling the entire IES, approaches are broadly categorized into two main types: holistic methods and component-based methods. The primary holistic technique is the Energy Hub (EH) method, while the Energy Bus (EB) method is the predominant component-based approach.

The EH method establishes coupling models to represent energy conversion between different carriers. It is an important framework for IES analysis.

Specifically, an EH is conceived as a conceptual unit that characterizes the fundamental input, output, conversion, and storage functions for various energy types. A generalized EH unit model can be formulated through an energy conversion relationship matrix. This method offers high versatility and scalability, abstracting complex multi-energy supply and demand characteristics into a simplified balance of energy input and output.

Despite its benefits, the EH method has limitations: critical information regarding energy flow parameters can be lost during the abstraction process. Moreover, by simplifying the system into a black box, the model may fail to adequately capture the internal dynamics of specific energy linkages or the system's multi-time scale operational characteristics.

The EB method is a technique that distinctly models different types of equipment and the associated energy flows. It constructs the IES mathematical model by setting up energy balance equations at various EB nodes, which act as connectors between individual equipment models. The bus-based structure provides an intuitive representation of the connections and coupling mechanisms among different devices. It also clearly depicts the conversion paths between energy media, making it a widely adopted method for modeling regional IES. The effectiveness of any EB-based system model is directly contingent upon the accuracy of the underlying internal equipment models.

Modeling techniques for individual equipment within the IES are typically divided into three categories, based on the required level of understanding of the equipment's internal structure and function:

White-Box Method: This is a mechanistic modeling approach requiring a detailed description of the equipment's internal processes and physical reactions. While data-calibrated white-box models offer high accuracy and are excellent for analyzing system operational characteristics, they necessitate numerous parameters and often involve long computation times.

Black-Box Method: This approach constructs a model of outputs solely based on inputs, without incorporating any physical knowledge of the internal mechanism. Artificial neural networks are the most common implementation of this method.

Gray-Box Method: This method strategically combines mechanism-based knowledge and data-driven techniques. Gray-box models are generally simpler and faster to compute, provide a degree of empirical insight, and are interpretable. The most prevalent gray-box approach in IES modeling is the simplified coefficient method.

This chapter will predominantly employ the EB method. It will establish foundational models for typical IES equipment, covering the source, load, and storage perspectives, thereby creating the necessary model framework for subsequent optimization and scheduling research.

3.1.2 General system modeling method

This section describes the energy conversion characteristics of common equipment in IES, starting from the source, load, and storage sides. This includes primary energy conversion equipment like combined heat and power (CHP) units, secondary energy conversion equipment like electric boilers, user demand, and energy storage devices like batteries. To facilitate the rapid modeling of various types of energy equipment that may be added to the IES in the future, this section, inspired by EH and computer modeling methods, proposes a component-based modeling approach. This method categorizes equipment models into source-side energy conversion, storage, and load types, using a consistent model structure to represent each equipment category. These models are then interconnected through energy balance equations at various EBs to build an integrated model of the IES.

3.1.2.1 Source-side equipment modeling

Source-side equipment primarily performs energy conversion functions. Although the principles of different energy conversion devices vary, they can all be abstracted into an input-output relationship matrix, which can be either a numerical or a non-linear matrix. In the operational scheduling decisions of an IES, the inputs and outputs of each source-side device need to be optimized. This chapter refers to source-side energy conversion equipment as the component "U". Let the input energy flow be matrix X, the conversion matrix be C, and the bias matrix be B, then the output matrix Y can be calculated by Y = CX + B.

3.1.2.1.1 Coal-Fired Boiler

A coal-fired boiler converts coal into steam, requiring control of coal and air flow parameters to ensure safety and produce the required amount of steam. Its mathematical model can be described as follows:

$$P_{\mathrm{CB},t} = v_{\mathrm{coal},t} \times H_{\mathrm{coal},t} \times \eta_{\mathrm{CB},t} \tag{3.1}$$

where $P_{\mathrm{CB},t}$ is the heat supply of the coal-fired boiler at time t, $v_{\mathrm{coal},t}$ is the coal consumption at time t, $H_{\mathrm{coal},t}$ is the calorific value of coal at time t, and $\eta_{\mathrm{CB},t}$ is the thermal efficiency of the boiler at time t.

3.1.2.1.2 Gas-fired hot water boiler

A gas-fired hot water boiler uses natural gas as fuel. The high-temperature steam it produces passes through a heat exchanger to supply hot water. Its mathematical model is as follows:

$$P_{GB,t} = v_{GB,t} \times H_{CH4,t} \times \eta_{GB,t} \tag{3.2}$$

$$\theta_{GB,out,t} = P_{GB,t} / \left(c_{GB,t} \times v_{GB,H2O,t} \right) + \theta_{GB,in,t} \tag{3.3}$$

where $P_{GB,t}$ is the heat supply at time t, $v_{GB,t}$ is the natural gas consumption at time t, $H_{CH4,t}$ is the calorific value of natural gas at time t, $\eta_{GB,t}$ is the thermal efficiency at time t, $\theta_{GB,out,t}$ is the inlet water temperature, $\theta_{GB,in,t}$ is the outlet water temperature, $c_{GB,t}$ is the average specific heat capacity of water, and $v_{GB,H2O,t}$ is the mass flow rate of water.

3.1.2.1.3 Wind turbine generator

Wind power generation converts the kinetic energy of wind into mechanical energy, and then into electrical energy. Due to the uncertainty and volatility of wind speed, the power output of a wind turbine is also random and fluctuating. The mathematical model is generally a piecewise function of wind speed:

$$P_{WP,t} = \begin{cases} 0, & 0 \leq \upsilon_{WP,t} < \upsilon_{WP,in}, \upsilon_{WP,t} > \upsilon_{WP,out} \\ \dfrac{\upsilon_{WP,t} - \upsilon_{WP,in}}{\upsilon_{WP,rated} - \upsilon_{WP,in}} \times P_{WP,rated}, & \upsilon_{WP,in} \leq \upsilon_{WP,t} < \upsilon_{WP,rated} \\ P_{WP,rated} & \upsilon_{WP,rated} \leq \upsilon_{WP,t} \leq \upsilon_{WP,out} \end{cases} \tag{3.4}$$

where $\upsilon_{WP,t}$ is the wind speed at time t, $\upsilon_{WP,in}$ is the cut-in speed, $\upsilon_{WP,out}$ is the cut-out speed, $\upsilon_{WP,rated}$ is the rated speed, and $P_{WP,rated}$ is the rated power of the wind turbine.

3.1.2.1.4 Photovoltaic (PV) generator

A PV generator converts light energy into electrical energy using the PV effect. The mathematical model is as follows:

$$P_{PV,t} = P_{PV,rated} \times \frac{G_{PV,t}}{G_{PV,rated}} \left[1 + k_{PV} \left(\theta_{PV,t} - \theta_{PV,rated} \right) \right] \tag{3.5}$$

where $P_{PV,t}$ is the power output at time t, $P_{PV,rated}$ is the rated power under standard conditions, $G_{PV,t}$ is the actual solar irradiance at time t, $G_{PV,rated}$ is the standard irradiance, k_{PV} is the power temperature coefficient, $\theta_{PV,t}$ is the actual temperature at time t, and $\theta_{PV,rated}$ is the standard temperature.

3.1.2.1.5 Electric boiler

An electric boiler converts electrical energy into thermal energy. Integrating it into a CHP system can decouple the "heat-determines-power" mode, thereby enhancing system flexibility and renewable energy accommodation. Its mathematical model is as follows:

$$P_{EB,t} = P_{EB,in,t} \times \eta_{EB} \tag{3.6}$$

where $P_{EB,t}$ is the heat supply at time t, $P_{EB,in,t}$ is the power consumption at time t, and η_{EB} is the electro-thermal conversion efficiency.

3.1.2.1.6 Steam turbine

A steam turbine converts the thermal energy of high-temperature, high-pressure steam from a boiler into mechanical energy, which then drives a generator. Its mathematical model is as follows:

$$P_{TP,t} = a_{TP,0} \times v_{TP,in,t} + a_{TP,1} \times v_{TP,out0,t} + a_{TP,2} \times v_{TP,out1,t} + a_{TP,3} \tag{3.7}$$

where $P_{TP,t}$ is the power generation at time t, $v_{TP,in,t}$ is the inlet steam flow, $v_{TP,out0,t}$ is the extraction steam flow, $v_{TP,out1,t}$ is the exhaust steam flow, and $a_{TP,0}$, $a_{TP,1}$, $a_{TP,2}$ are the turbine's power generation model coefficients.

3.1.2.1.7 CHP unit

CHP is an efficient form of energy conversion that transforms a single energy source into both electricity and heat. Using a condensing-extraction coal-fired CHP unit as an example, its mathematical model can be described as follows:

$$v_{coal,t} = v_{bs,t} \times H_{b,t} / \left(H_{coal,t} \times \eta_{b,t} \right) \tag{3.8}$$

$$P_{CHP,t} = a_{CHP,0} \times v_{TP,in,t} + a_{CHP,1} \times v_{TP,out0,t} + a_{CHP,2} \times v_{TP,out1,t} + a_{CHP,3} \tag{3.9}$$

where $v_{coal,t}$ is the coal consumption of the boiler at time t, $v_{bs,t}$ is the steam production, $H_{b,t}$ and $H_{coal,t}$ are the enthalpies of steam and feedwater, $\eta_{b,t}$ is the boiler efficiency, $P_{CHP,t}$ is the power generation, and other terms are as defined for the steam turbine.

3.1.2.1.8 Power-to-gas (P2G) equipment

P2G equipment converts electricity to natural gas through two steps: water electrolysis and methanation. Its mathematical model is as follows:

$$P_{P2G,t} = P_{P2G,in,t} \times \eta_{P2G,G} \tag{3.10}$$

$$P_{P2G,out,t} = P_{P2G,in,t} \times \eta_{P2G,H} \tag{3.11}$$

where $P_{P2G,t}$ is the natural gas production power at time t, $P_{P2G,in,t}$ is the electricity consumption, $\eta_{P2G,G}$ is the gas production efficiency, $\eta_{P2G,G}$ is the heat supply, and $\eta_{P2G,H}$ is the heat production efficiency.

3.1.2.1.9 Heat pump

A heat pump transfers heat from a low-temperature source to a high-temperature sink. For operational decisions in an IES, its model can be simplified to an electro-thermal conversion model:

$$P_{HP,t} = P_{HP,in,t} \times \eta_{HP} \tag{3.12}$$

where $P_{HP,t}$ is the heat supply at time t, $P_{HP,in,t}$ is the power consumption, and η_{HP} is the heating efficiency.

3.1.2.2 Load-side forecasting and equipment modeling

Loads can be classified as rigid or flexible. The load types downstream of an IES are diverse. This chapter employs data-driven methods for forecasting industrial steam and gas loads, while using a unified mathematical model for load-side equipment. Load-type equipment is designated as component "D". The input X and output Y both consist of flow rates of various energy carriers. Here, Y represents the demand value, and X represents the actual energy supplied.

3.1.2.2.1 Industrial steam and gas load forecasting

The energy demand of industrial users varies with their production schedules, making it unstable and irregular. It is influenced by various factors such as weather, economy, and policy, making it difficult to represent with mechanistic models. Accurately forecasting industrial user demand is key to achieving efficient operational scheduling for an IES. This chapter uses a hybrid forecasting method based on features and time series to improve accuracy.

Using a Long Short-Term Memory (LSTM) network, a short-term load forecasting model is established. The input feature parameter X is a collection of temperature, pressure, quantity, and time parameters over a past period.

$$X = \left[p, v, \theta, x_m, x_d, x_h \right] \tag{3.13}$$

$$\begin{cases} x_m = x_m / 12 \\ x_d = x_d / 31 \\ x_h = x_h / 24 \\ v_{\text{pre}} = f(X) \end{cases} \tag{3.14}$$

where $p, v, \theta, x_m, x_d, x_h$. are pressure, quantity, temperature, and time features; X is the LSTM model input; v_{pre} is the forecasted load for the future period; and f is the trained LSTM-based forecasting model.

3.1.2.2.2 High-energy-consuming electrical load modeling

Flexible high-energy-consuming loads can be divided into discretely adjustable, transferable, and continuously adjustable loads.

Discretely Adjustable Load: Can be increased or decreased by a fixed amount within a specified time but requires a certain stabilization period.

$$P_{\text{ld},t} = P_{\text{ld},t-1} - \left(1 - x_{\text{ld},t} \right) \times d_{\text{ld},t} \tag{3.15}$$

where $P_{\text{ld},t}$ is the load value at time t; $x_{\text{ld},t}$ is a binary variable indicating if the load participates in regulation (0 for yes, 1 for no); and $d_{\text{ld},t}$ is the downward adjustment amount.

Transferable Load: Can shift its demand to other time slots while keeping the total load within a certain period constant, thus reducing peak load.

$$P_{\text{lt},t} = P_{\text{lt},t-1} + x_{\text{lt},t} \times d_{\text{lt},t} \tag{3.16}$$

where $P_{\text{lt},t}$ is the load value at time t and $d_{\text{lt},t}$ is the load shifted to time t.

Continuously Adjustable Load: Can respond to grid commands continuously and rapidly in each period without needing stabilization time.

$$P_{\text{lc},t} = \sum_{n=1}^{n=N_{\text{lc}}} d_{\text{lc},n,t} \tag{3.17}$$

where $P_{lc,t}$ is the total continuously adjustable load at time t and $d_{lc,n,t}$ is the output of the nth continuously adjustable unit.

3.1.2.3 Storage-side equipment modeling

Storage-side equipment mainly includes conventional single-energy storage devices like batteries, thermal storage tanks, and gas holders, as well as multi-energy storage devices like Carnot batteries. We designate single-energy storage devices as component "C". The input X represents charging energy, and Y represents discharging energy. Additional matrices $P_{s,in}$, $P_{s,out}$, $P_{s,init}$, and P_s represent charging efficiency, discharging efficiency, initial stored energy, and current stored energy, respectively. The relationship between them is as follows:

$$P_s = P_{s,init} + X \times P_{s,in} - \frac{Y}{P_{s,out}} \tag{3.18}$$

3.1.2.3.1 Battery

$$P_{se,t} = P_{se,t-1} + P_{se,in,t} \times \eta_{se,in} - \frac{P_{se,out,t}}{\eta_{se,out}} \tag{3.19}$$

where $P_{se,t}$ is the stored energy at time t, $P_{se,in,t}$ is the charging power, $P_{se,out,t}$ is the discharging power, and $\eta_{se,in}$ and $\eta_{se,out}$ are charging and discharging efficiencies.

3.1.2.3.2 Thermal storage tank/ice storage tank

Using a thermal storage tank as an example, the operational process model is as follows:

$$P_{sh,t} = P_{sh,t-1} + P_{sh,in,t} \times \eta_{sh,in} - \frac{P_{sh,out,t}}{\eta_{sh,out}} \tag{3.20}$$

where $P_{sh,t}$ is the stored heat at time t, $P_{sh,in,t}$ is the heat charged, $P_{sh,out,t}$ is the heat discharged, and $\eta_{sh,in}$ and $\eta_{sh,out}$ are charging and discharging efficiencies.

3.1.2.3.3 Gas holder

$$P_{sg,t} = P_{sg,t-1} + P_{sg,in,t} \times \eta_{sg,in} - \frac{P_{sg,out,t}}{\eta_{sg,out}} \tag{3.21}$$

where $P_{sg,t}$ is the stored gas volume at time t, $P_{sg,in,t}$ is the gas charged, $P_{sg,out,t}$ is the gas discharged, and $\eta_{sg,in}$ and $\eta_{sg,out}$ are efficiencies.

3.1.2.3.4 Carnot battery

A Carnot battery is an emerging energy storage technology. It consists of a compressor, a turbine, and two thermal storage units, capable of converting electricity into heat for storage. When electricity demand is high, it can convert the stored heat back into electricity and can also supply heat directly. Its modeling approach is as follows:

$$P_{\text{cbp2h},t} = a_{\text{cb}} \times P_{\text{cb,in},t} \tag{3.22}$$

$$P_{\text{cb,out},t} = COP_{\text{cb}} \times P_{\text{cbh2p},t} \tag{3.23}$$

$$P_{\text{cb},t} = P_{\text{cb},t-1} + P_{\text{cbp2h},t} \times \eta_{\text{cb,in}} - \frac{P_{\text{cbh2p},t} + P_{\text{cbho},t}}{\eta_{\text{cb,out}}} \tag{3.24}$$

where $P_{\text{cbp2h},t}$ is the heat generated from electricity at time t, a_{cb} is the power-to-heat coefficient, $P_{\text{cb,in},t}$ is the charging power, $P_{\text{cb,out},t}$ is the discharging power, COP_{cb} is the heat-to-power coefficient, $P_{\text{cbh2p},t}$ is the heat converted to power, $P_{\text{cb},t}$ is the stored heat, $\eta_{\text{cb,in}}$ is the storage efficiency, and $P_{\text{cbho},t}$ is the direct heat supply.

3.2 OPTIMIZATION SCHEDULING TECHNOLOGIES FOR IES

3.2.1 IES scheduling framework

A typical optimal scheduling framework for an Integrated Electricity and Heating System. The objective of this framework is to minimize the total operating cost of the system while meeting both electricity and heat load demands and satisfying all operational constraints.

Input Layer: This includes forecasts for the next scheduling period (e.g., 24 h) for electricity and heat loads, renewable energy (e.g., wind, solar) output, as well as economic parameters such as fuel (coal, natural gas) prices and electricity prices.

Decision Layer: This is the core of the scheduling process. The optimization model, based on the input information, determines the operational plan for all controllable devices in the system. The decision variables mainly include:

1. The on/off status and power output of conventional and CHP generating units.
2. The heat production of CHP units and gas boilers.
3. The charging/discharging power of electricity and heat storage devices.
4. The curtailment amount of interruptible loads.

Constraint Layer: This ensures the feasibility and security of the decision plan. It includes all the equality and inequality constraints mentioned earlier, especially the accurately modeled power grid flow constraints and heating network dynamic constraints.

Output Layer: The final optimal scheduling plan, which is a detailed operational curve for each device over the future scheduling period.

3.2.2 Scheduling based on traditional intelligent optimization algorithms

For solving the complex MINLP problem of IES scheduling, when the model can be effectively simplified or linearized, commercial solvers (such as Gurobi, CPLEX) can be used for exact solutions. However, when the model is highly nonlinear and non-convex, or when the problem scale is very large, traditional mathematical programming methods may face issues of long computation times or even failure to converge. In such cases, heuristic intelligent optimization algorithms demonstrate their advantages.

Heuristic algorithms, such as Genetic Algorithms (GA), Particle Swarm Optimization (PSO), and Simulated Annealing, are a class of stochastic search algorithms that mimic natural biological evolution or physical processes. They do not rely on the gradient information of the problem and have no requirements for the continuity or convexity of the objective function, thus possessing strong robustness and global search capabilities, making them suitable for solving complex, black-box optimization problems.

Taking PSO as an example, its application process in IES scheduling is as follows:

1. Particle Initialization: A swarm of particles is generated, with each particle representing a potential scheduling solution. The position vector of a particle contains the values of all control variables over a scheduling period. For instance, for a 24-h schedule with N control variables, the dimension of the particle's position vector would be

24 × N. The position and velocity of each particle are initialized randomly.

2. Fitness Evaluation: For each particle (i.e., each scheduling plan), it is first checked whether it satisfies all hard constraints (e.g., power balance, equipment output limits). For infeasible solutions, a large penalty can be applied to their fitness value using a penalty function method, so that they are eliminated during the evolutionary process. For feasible solutions, the objective function value (e.g., total operating cost) is calculated as its fitness. The lower the cost, the higher the fitness.

3. Update Personal and Global Best: Each particle records its own best-experienced position (pBest). Simultaneously, the entire swarm records the best position experienced by any particle (gBest).

4. Update Particle Velocity and Position: This is the core of the optimization. Each particle adjusts its current schedule based on a combination of three influences, much like a bird in a flock adjusting its flight path: (a) Its own momentum: It tends to continue adjusting its schedule in the same direction it was previously. (b) Its own experience: It is pulled toward its own personal best solution. (c) The group's experience: It is also pulled toward the swarm's global best solution. By combining these tendencies with a slight amount of randomness, each particle explores new, potentially better scheduling possibilities. It learns from its own successes and from the successes of the entire group.

5. Termination Condition Check: The process is concluded when the specified criteria for stopping are satisfied. These conditions typically involve reaching a predetermined maximum count of iterations or achieving the required precision in the solution. The gBest achieved at the end of the process represents the most favorable scheduling solution.

The GA is a stochastic search technique well-suited for addressing complex optimization challenges that are non-linear, feature multiple constraints, and simulate the mechanisms of natural evolution. The principal stages of the GA are outlined below:

1. Encoding and Initialization: A population of initial solutions is randomly generated. Each individual (chromosome) in the population represents a potential scheduling plan, consisting of the decision variables to be optimized, such as the thermal output of CHP units, thermal power units, gas boilers, and the output of electric air compressors.

2. Fitness Evaluation: The fitness of each individual in the parent population is calculated based on the optimization objective function (e.g., total operating cost). A higher fitness value indicates a better economic benefit for the scheduling plan.

3. Selection: Individuals are selected from the parent population to form the next generation based on their fitness values. Individuals with higher fitness have a greater probability of being selected. Common selection strategies include roulette wheel selection and tournament selection.

4. Crossover: This step mimics the recombination of genes in biological reproduction. Pairs of individuals are randomly selected from the chosen population, and parts of their genes (decision variables) are exchanged at a certain crossover probability to generate new individuals (offspring).

5. Mutation: This step simulates genetic mutation in evolution. Some genes in the offspring individuals are randomly altered with a small mutation probability to increase population diversity and help escape local optima.

6. Formation of a New Population: The individuals resulting from the selection, crossover, and mutation operations form a new offspring population.

7. Termination Condition: Steps (2) to (6) are repeated until a termination condition is met, such as reaching the maximum number of iterations or the convergence of the population's fitness value. The individual with the highest fitness in the final population is the optimal scheduling plan.

Advantages:

1. The algorithm's principle is simple and easy to implement.
2. It has strong search capabilities and can effectively find high-quality approximate optimal solutions for non-convex, nonlinear problems.
3. It can conveniently handle multi-objective optimization problems (e.g., NSGA-II).

Disadvantages:

1. It is prone to getting stuck in local optima, especially in high-dimensional, complex problems.
2. The convergence speed of the algorithm is relatively slow, and its performance is sensitive to parameter settings (such as population size, learning factors), which require empirical tuning.
3. It cannot guarantee finding the global optimal solution.

3.2.3 Scheduling based on model predictive control (MPC)

MPC is an advanced control strategy that originated in the field of process control and is particularly well-suited for handling systems with complex dynamics, multiple variables, and constraints. Its core idea is to use a dynamic model of a system to predict its behavior over a finite future time range (the Prediction Horizon) at the current moment. An open-loop optimization problem is then solved to obtain a series of future control actions. However, only the first of this series of actions is applied to the system. At the next moment, the system state is updated based on new measurements, and the entire process is repeated. This rolling optimization approach creates a closed-loop feedback control, enabling it to effectively deal with disturbances and model uncertainties.

Application Framework of MPC in IES Scheduling:

1. State Measurement/Estimation: At each scheduling instant k (e.g., every 15 min), the current state of the system is obtained. This includes the state of charge of energy storage devices, the temperatures at key nodes in the heating network, and the latest ultra-short-term forecasts for loads and renewable energy output.
2. Rolling Optimization:

   Prediction: Based on the system's dynamic model (including equipment and network models) and the updated ultra-short-term forecasts (e.g., for the next 4–6 h of load and wind/solar generation), the dynamic response of the system over the next several time steps (the Prediction Horizon) is predicted.

   Optimization: An optimization problem is solved over the prediction horizon to minimize the total cost (or other performance metrics) within that timeframe. This optimization problem is similar to the scheduling problem described earlier, but its time window is rolling.

   Control Sequence: Solving this optimization problem yields an optimal control sequence for the next several time steps (the Control Horizon).
3. Control Implementation: Only the first element of the optimal control sequence is applied to the actual system.
4. Horizon Shift: Wait for the next scheduling instant $k+1$ to arrive, then return to step 1 to begin a new round of prediction and optimization.

Advantages of MPC:

1. Ability to Handle Constraints: MPC can directly and explicitly handle various equality and inequality constraints during the optimization process, ensuring that the scheduling plan is always safe and feasible.

2. Feedback Correction Mechanism: Through rolling optimization and state feedback, MPC can effectively cope with uncertainties and disturbances. When actual loads or renewable energy outputs deviate from forecasts, MPC can promptly adjust the control strategy at the next instant, exhibiting strong robustness.

3. Handling of Dynamic Coupling: MPC is very suitable for dealing with systems with complex dynamic characteristics, such as the delay effects in the heating network of an IES. It can accurately describe such dynamic processes in its prediction model, thereby formulating more forward-looking optimization strategies.

4. Multi-Objective Coordination: Multiple performance indicators (such as economy, environmental protection, stability) can be conveniently included in the objective function and coordinated through weights.

Challenges:

1. High Requirement for Model Accuracy: The performance of MPC is highly dependent on the accuracy of the prediction model. A mismatched model can lead to suboptimal or even unstable control effects.

2. Computational Burden: A complex optimization problem needs to be solved at each scheduling instant, which places high demands on computational resources. Ensuring the timeliness of online calculations for large-scale IES is a challenge. Efficient numerical algorithms or model simplification techniques need to be developed.

Summary and Comparison:

1. Traditional intelligent optimization algorithms are more like an "offline" global planning method. They attempt to find the optimal solution for the entire scheduling period (e.g., 24 h) at once. They are relatively less demanding in terms of model accuracy but have poor robustness to disturbances, functioning as an open-loop control.

2. MPC is an "online" rolling decision-making method. It continuously revises future plans based on the latest information, featuring a closed-loop feedback mechanism that is highly adaptive to uncertainties and disturbances. However, it relies on the accuracy of the model and online computational capability.

In practical applications, these two methods can also be combined. For example, a GA or a similar intelligent algorithm can be used for a rough day-ahead scheduling to provide a general plan and reserve capacity for the system.

Then, during the intra-day or real-time stage, MPC can be employed for rolling adjustments to cope with real-time changes and disturbances, thereby achieving a balance of economy, security, and stability.

3.3 MULTI-SCALE SCHEDULING STRATEGIES FOR IES

3.3.1 Impact of time granularity on system scheduling

Different energy forms exhibit unique properties across various time scales, which presents both challenges and opportunities for the development of system scheduling strategies.

3.3.1.1 Multi-time scale differences in energy characteristics and scheduling strategies

The dynamic properties of various energy carriers exhibit significant differences. Concurrently, the responses of equipment control systems and overall system demand to dispatch commands also show substantial inconsistencies. These factors collectively increase the complexity of optimizing operational scheduling. To effectively conduct the operational scheduling of an IES, it is imperative to consider the characteristic differences of various energy forms across temporal scales. In this context, the concept of "minimum time granularity" has been introduced. The selection of time granularity necessitates a trade-off among economic efficiency, uncertainty, and computational tractability. From an economic standpoint, a coarser time granularity allows for a better characterization of the coupled variations among multiple energy flows. From the perspective of uncertainty, a finer time granularity can mitigate the impacts of unforeseen fluctuations. From a computational efficiency viewpoint, a finer time granularity corresponds to a model with fewer variables and iterations, thereby enhancing computational speed. This section will analyze the influence of time granularity on the scheduling of IES from two dimensions to elucidate the importance of selecting an appropriate time granularity.

Different forms of energy exhibit unique properties on the time scale. Figure 3.1 illustrates the discrepancies among different energy carriers across various time scales.

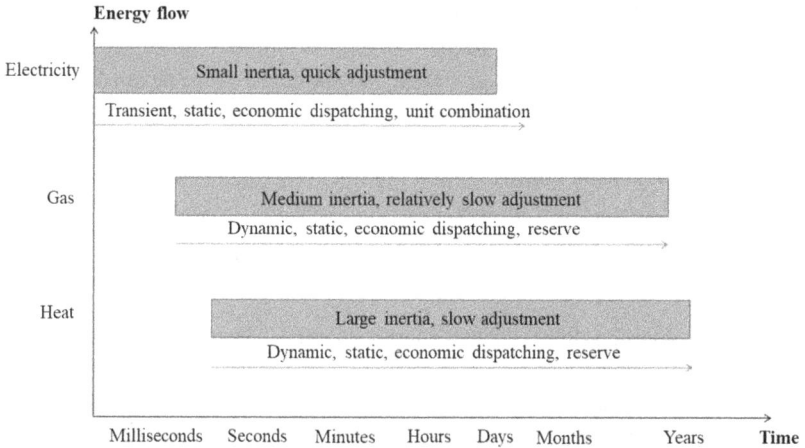

FIGURE 3.1 Time scale characteristics of different energy carriers.

This diagram mainly describes the time inertia characteristics of electricity, natural gas, and heat. Electricity has small inertia and fast regulation, natural gas has large inertia and slow regulation, and heat has large inertia and slow regulation.

Electricity is characterized by low inertia and rapid regulation, enabling responses at the millisecond level. Natural gas possesses greater inertia than electricity, resulting in a slower regulation speed. Thermal energy exhibits even higher inertia, with significant time lags in its adjustment. With respect to the time axis, for the electrical grid, the millisecond scale pertains to transient stability, whereas on scales of seconds or longer, the system can be considered in a steady state. In the context of economic dispatch and unit commitment, electricity is often treated as a non-deferrable (rigid) energy source, corresponding to time scales of minutes, hours, days, or months. For natural gas, its dynamic processes are most prominent at the millisecond-to-minute level. For thermal systems, the thermal dynamics are most apparent on scales of tens of seconds to minutes. When considering the vertical axis (energy type), scheduling different energy forms along the same timeline requires accounting for their distinct properties. For instance, on second- and minute-level time scales, only electricity can be regarded as a rigid and rapidly responding energy source.

In addition to the intrinsic time scale differences of energy carriers, current scheduling strategies also encompass multiple time scales, typically including day-ahead, intra-day, and real-time optimization. The day-ahead schedule is formulated with an hourly time granularity for the subsequent 24-h period. The intra-day schedule, which refines the hourly day-ahead plan, operates on an hourly or minute-level granularity, updating the schedule from the present moment until the

24-h mark based on more detailed forecasts of renewable energy and demand. The real-time dispatch, which further refines the intra-day schedule, utilizes a 15-min, 5-min, or 1-min time granularity to formulate more detailed dispatch instructions. Through this rolling-horizon correction methodology across different time scales, the impact of uncertainties originating from renewable energy generation and demand forecasting can be effectively diminished.

3.3.1.2 Multi-time scale matching characteristics of supply and demand

The aforementioned differences in energy properties lead to significant variations in the dynamic characteristics of equipment within an IES, which allows for a temporal offset between supply and demand. Concurrently, the fulfillment of different energy demands also varies, providing a degree of flexibility for scheduling. The thermal inertia of buildings, the heat storage capacity of district heating networks, and the line-pack capability (compressibility) of natural gas pipelines can all serve as flexibility resources, creating an elastic buffer between supply and demand. As depicted in Figure 3.2, supply and demand exhibit characteristics of asynchronous matching across multiple time scales. In traditional models, supply and demand curves are aligned at identical time points. However, when multi-energy flexibility resources are considered, both supply and demand possess a range of flexibility. Moreover, supply and demand may not align at the same time instances, manifesting as a discrepancy in the trends of their respective curves. Through the complementarity of energy flows, demand shifting and conversion can be realized. For instance, an

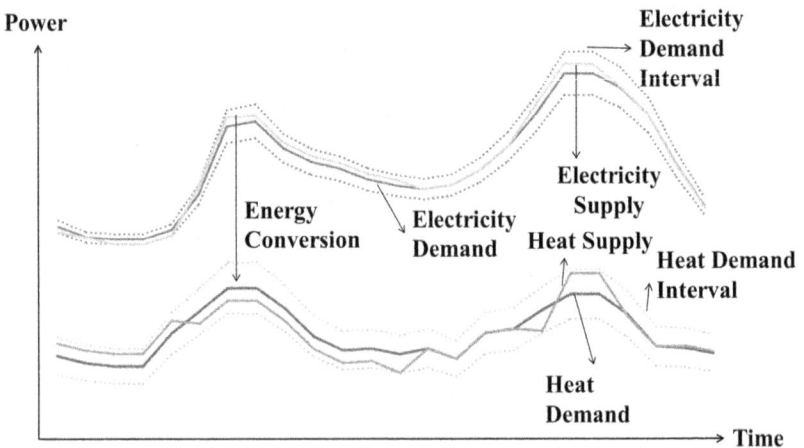

FIGURE 3.2   Multi-time scale matching characteristics of supply and demand.

increase in electricity supply can be leveraged by power-to-heat technologies to satisfy thermal demand.

This graph takes the electricity and heat loads as examples to describe the supply and demand balance relationship on the time series curve.

In addition to the multi-scale matching of supply and demand, energy efficiency and economic benefits vary with different time granularities. From an economic perspective, the optimization process benefits from a coarser time granularity to better capture the coupling dynamics among the system's various energy flows, thereby fully exploiting energy flow complementarity and the functionality of energy storage systems. From the perspective of robustness against disturbances, a finer time granularity is preferable to reduce the impact of system uncertainties, as short-term forecasts for renewable generation and demand are more accurate than long-term ones.

In an IES, the efficiency of equipment is contingent upon its instantaneous operating state and load demand. Therefore, from an energy efficiency standpoint, selecting a finer time granularity enables the achievement of higher efficiency at a more detailed operational level, consequently improving the average overall efficiency.

3.3.2 Multi-time scale flexible scheduling model based on a unified time granularity

The flexibility resources in multi-energy supply and demand create a wider optimization space for scheduling. Simultaneously, the multi-scale characteristics of energy carriers, scheduling strategies, and supply-demand matching present both opportunities and challenges for the operational scheduling of an IES. Significant differences exist in energy efficiency, economic benefits, uncertainty, and computational tractability at varying time granularities. Thus, the selection of an appropriate time granularity that holistically considers these factors is paramount for achieving the flexible scheduling of multiple energy forms across different time scales.

This section addresses the ultra-short-term scheduling optimization problem for an IES by establishing a multi-time scale dynamic optimization model that considers environmental, economic, and feasibility criteria. The scheduling horizon of this model is 1 h. The decision inputs consist of the forecasted multi-energy demand sequences for the upcoming hour, which can be adjusted based on prices, incentives, and network inertia. The decision outputs comprise the load allocation for each piece of equipment within the system for the next hour, subject to constraints such as equipment capacity, user demand, and renewable energy generation. The objective is to maximize

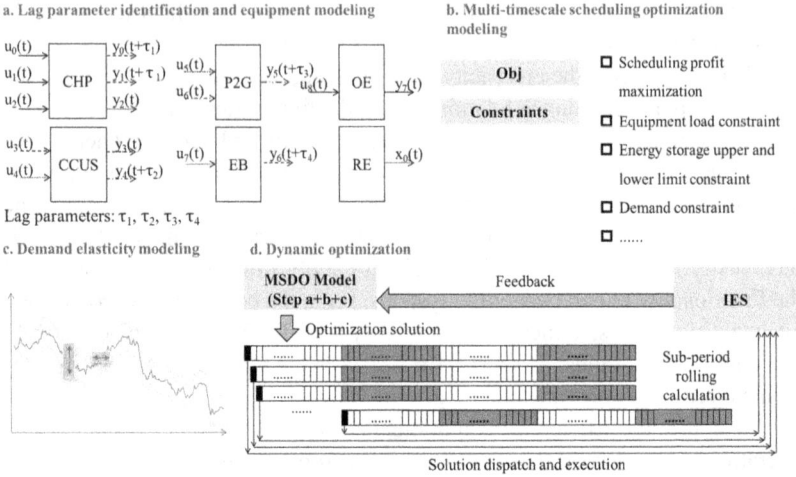

FIGURE 3.3   Flowchart of the multi-time scale dynamic optimization methodology.

the overall economic benefit while giving full consideration to environmental performance. Specifically, Figure 3.3 presents the workflow of the proposed multi-time scale dynamic optimization methodology.

(This figure depicts a simulation framework for multi-time scale dynamic optimization, which includes four steps: lag parameter identification and equipment modeling, multi-time scale scheduling optimization modeling, demand elasticity modeling, and dynamic optimization.)

The methodology comprises four principal components:

1. Lag Parameter Identification and Equipment Modeling: Correlation analysis is employed to identify the response time differentials among various pieces of equipment, and data-fitting techniques are used to establish their respective input-output relationship models.
2. Multi-time Scale Scheduling Optimization Modeling: Based on the identified equipment models and accounting for their multi-scale response differences, a suitable minimum time granularity is selected. An optimization model is then formulated with the objective of maximizing the overall economic benefit of the IES, subject to operational constraints such as equipment performance and user demand, to determine the optimal scheduling strategy.
3. Demand Elasticity Modeling: The demand constraints from the preceding modeling step are augmented to include upper and lower bounds. Constraints related to demand response, cumulative demand, and energy network inertia are also formulated.

4. Dynamic Optimization: Leveraging the multi-time scale scheduling model, the flexible matching of supply and demand at different time instances is achieved by considering bilateral flexibility. The schedule for subsequent periods is refined based on actual execution data until the first sub-period concludes. Subsequently, updated demand forecasts are acquired, and the multi-period optimization is re-executed.

3.3.2.1 Objective function

The objective of the multi-time scale dynamic optimization for the IES is to maximize the overall benefit of the supply system. The overall economic income includes sales of steam, hot water, electricity, natural gas, and carbon emission rights, while expenditures include penalties for wind/solar curtailment, coal consumption costs, and penalties for unmet demand.

$$\max E_{\text{total}} = \sum_{t=0}^{T}\sum_{i=0}^{5} v_{\text{supply},i,t} \times \varphi_{i,t} - \sum_{t=0}^{T}\sum_{j=0}^{2} v_{\text{punish},j,t} \times \varphi_{j,t} \tag{3.25}$$

where E_{total} is the total benefit of the supply system; T is the entire scheduling period; $v_{\text{supply},i,t}$ is the quantity of energy type i sold at time t; $\varphi_{i,t}$ is the price of energy type i at time t; $v_{\text{punish},j,t}$ is the quantity of penalty item j (e.g., curtailed wind/solar, coal consumption) at time t; and $\varphi_{j,t}$ is the penalty cost coefficient for item j at time t.

3.3.2.2 Constraints

3.3.2.2.1 Equipment load upper and lower limit constraints

Due to equipment design and actual production capacity limitations, there are upper and lower limits on the energy consumed and produced by the equipment.

$$L_{\text{in},m,\text{lb}} \leq u_{m,t} \leq L_{\text{in},m,\text{ub}} \tag{3.26}$$

$$L_{\text{out},n,\text{lb}} \leq y_{n,t} \leq L_{\text{out},n,\text{ub}} \tag{3.27}$$

where $L_{\text{in},m,\text{lb}}$ is the lower limit of equipment energy consumption, $u_{m,t}$ is the amount of equipment energy consumption at time t, $L_{\text{in},m,\text{ub}}$ is the upper limit of equipment energy consumption, $L_{\text{out},n,\text{lb}}$ is the lower limit of equipment energy production, $y_{n,t}$ is the amount of equipment energy production at time t, and $L_{\text{out},n,\text{ub}}$ is the upper limit of equipment energy production.

3.3.2.2.2 Minimum load constraint when equipment is on

When the equipment is started, it needs to meet the minimum operating load requirement.

$$\text{if } u_{m,t} > 0 : u_{m,t} \geq L_{\text{in},m,\text{ub}} \times r_m \tag{3.28}$$

where r_m is the proportion of the minimum operating load to the maximum load for the equipment when it starts.

3.3.2.2.3 Equipment performance constraints

Equipment performance constraints refer to the input-output relationship model of the equipment.

3.3.2.2.4 Demand constraints

The external energy supply of the integrated energy supply system needs to meet the user's demand.

3.3.2.2.5 Energy storage equipment constraints

Energy storage devices have upper and lower limits on their storage capacity.

$$P_{\text{sx},k,t+1} = P_{\text{sx},k,t} + P_{\text{sx,in},k,t}\eta_{\text{sx,in},k} - \frac{P_{\text{sx,out},k,t}}{\eta_{\text{sx,out}}} \tag{3.29}$$

$$L_{\text{sx},k,\text{lb}} \leq P_{\text{sx},k,t} \leq L_{\text{sx},k,\text{ub}} \tag{3.30}$$

where $P_{\text{sx},k,t+1}$ is the energy storage of the energy storage device at time t, $P_{\text{sx,in},k,t}$ is the amount of energy charged into the energy storage device at time t, $P_{\text{sx,out},k,t}$ is the amount of energy discharged from the energy storage device at time t, $L_{\text{sx},k,\text{lb}}$ is the lower limit of the energy storage of the energy storage device, and $L_{\text{sx},k,\text{ub}}$ is the upper limit of the energy storage of the energy storage device.

3.3.2.2.6 Equipment ramping constraints

There is an upper limit on the amount of load adjustment for each piece of equipment.

$$\left| u_{m,t+1} - u_{m,t} \right| \leq L_{\text{demand},d,\text{ub},m} \tag{3.31}$$

where $L_{\text{demand},d,\text{ub},m}$ is the upper limit of the equipment's load ramping capability.

3.3.2.2.7 Node input and output energy conversion

The energy flow branch nodes need to satisfy the constraint that the input energy is equal to the output energy.

3.3.2.2.8 Initial value constraints

$$P_{sx,k,t} = P_{sx,k,\text{init}} \tag{3.32}$$

$$u_{m,t} = u_{m,\text{init}} \tag{3.33}$$

$$y_{n,0} = y_{n,\text{init}} \tag{3.34}$$

where $P_{sx,k,\text{init}}$ is the initial energy storage of the energy storage device, $u_{m,\text{init}}$ is the initial load of the equipment, and $y_{n,\text{init}}$ is the initial energy production of the equipment.

3.3.2.3 Online dynamic optimization strategy

During online dynamic optimization, the input parameters are the demand values for the upcoming hour. Using a 15-min time interval as the unit, this includes four sets of demand forecasts. The established multi-time scale dynamic optimization model for an integrated energy supply system, which accounts for demand elasticity, is invoked as shown in the following equation to calculate and output the internal optimal dispatch plan for the integrated energy supply system for the next hour.

$$P_{sx}, u, y = f\left(P_{sx,k,\text{init}}, u_{m,\text{init}}, y_{n,\text{init}}, v_{\text{demand},i,t_p}, W_{\text{para}}\right) \tag{3.35}$$

In the equation, f represents the constructed multi-time scale dynamic optimization (MSDO) model for the integrated energy supply system, while W_{para} represents the set of upper and lower limit parameters, and other model parameters for variables such as $L_{\text{in},m,\text{lb}}$, $L_{\text{in},m,\text{ub}}$, etc.

The optimal dispatch plan is divided into four periods, and each period is discretized into a finer time granularity. First, the plan for the initial time point of the first sub-period is sent to the integrated energy supply system for execution. Subsequently, actual execution data from the system is collected and fed back into the optimization model to update the plan for the first sub-period.

When the first sub-period is complete, the system receives the latest demand forecast data for the next four periods and recalculates the optimal plan for these four sub-periods accordingly. This process is cyclical: adjustments are made and executed only for the plan within the first sub-period until it is fully completed. Afterward, the optimization plan for the next four periods is updated.

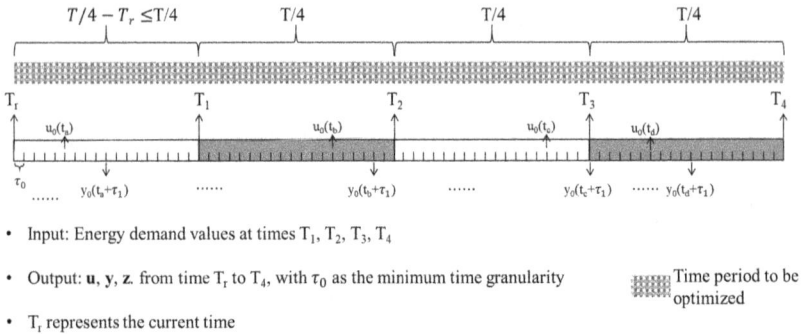

- Input: Energy demand values at times T_1, T_2, T_3, T_4

- Output: **u**, **y**, **z**. from time T_r to T_4, with τ_0 as the minimum time granularity

- T_r represents the current time

Time period to be optimized

FIGURE 3.4   Timeline of the MSDO method.

This differs from traditional dispatch optimization methods, where a dispatch decision at time t corresponds to the demand at time t. The unified time granularity method proposed in this chapter considers the differences in equipment response times, meaning a device input at time t may correspond to an output at time $t + k$. Moreover, the output at time $t + k$ does not need to strictly meet the demand at that exact moment, allowing for a degree of flexibility. This difference in time scales is what leads to the "dynamic" performance of the decision optimization.

Overall, this section employs a dynamic optimization method to ensure that cumulative demand constraints are met. The operational plan for the upcoming first sub-period is updated based on actual execution performance. The dispatch plan in this section features a fine time granularity and is achieved through single-level optimization. Figure 3.4 is a timeline diagram illustrating the MSDO method.

This figure mainly describes the dynamic optimization of multiple time scales, taking the energy demand at time T1, T2, T3, and T4 as input, to solve the start-stop state, power, and other variables of the equipment from the current moment to time T4 under the minimum time granularity.

3.4 CASE STUDY

3.4.1 Case settings

The IES selected in the examples of this chapter originates from a specific industrial park, which is equipped with four sets of 220 t/h high-temperature and ultra-high-pressure circulating fluidized bed boilers, two sets of 30 MW extraction back-pressure steam turbine generator sets, and two sets of 1500 Nm³/min back-pressure steam turbine compressed air units. The park's primary function is to provide energy for printing and dyeing enterprises, with steam being the predominant energy source for these industries. The park caters to 33 medium-pressure steam users and 211 low-pressure steam users, with an annual heating capacity of 19.629 million tons. The analysis of the data from this specific operating environment enables the study to deduce the operation scheduling strategy of an integrated energy supply system with similar configurations and supply and demand patterns (Figure 3.5).

This diagram depicts the framework diagram of the IES in the case study. The input energy sources include wind power, PV power, and coal. The terminal energy demands are electricity, heat, steam, natural gas, and carbon dioxide loads. The intermediate energy conversion equipment includes boilers, cogeneration units, carbon capture equipment, electric-to-gas equipment, electric boilers, etc. The energy storage equipment includes batteries and heat storage devices.

The small time interval is set to a granularity of 1 min, each sub-cycle is set to span 15 min, and the total cycle length is set to 1 h. The energy demand and equipment time lag parameters are configured according to Tables 3.1 and 3.2, and given the differences in energy characteristics, parameters such as the opening degree of the steam turbine inlet valve have been added. For specific parameter settings, please refer to Table 3.3.

3.4.2 Analysis of multi-time scale optimized scheduling results

The demand elasticity space is defined as the adjustment amount that fluctuates up and down in accordance with demand. The optimization calculation was conducted with the demand elasticity space set to 0, and the results are presented in Table 3.4. This table lists the energy supply and abandoned wind and solar values after optimization using the method outlined in this chapter. As illustrated in the table, the symbols $T_1, T_2, T_3,$ and T_4 are used to denote four distinct time points in relation to demand. A comparison with Table 3.1 reveals that, in the absence of demand elasticity, the optimization method developed in this chapter is capable of meeting the demands of low-pressure steam, medium-pressure steam, and electricity at each demand time point. In the context of non-strictly demanded hot water supply, the supply of carbon dioxide and natural gas is greater than zero. In addition, the abandoned wind and solar power values at the four demand time points are all zero, indicating that the optimization method adopted in this chapter can effectively achieve the complete consumption of renewable energy through the optimal scheduling of the integrated energy supply system.

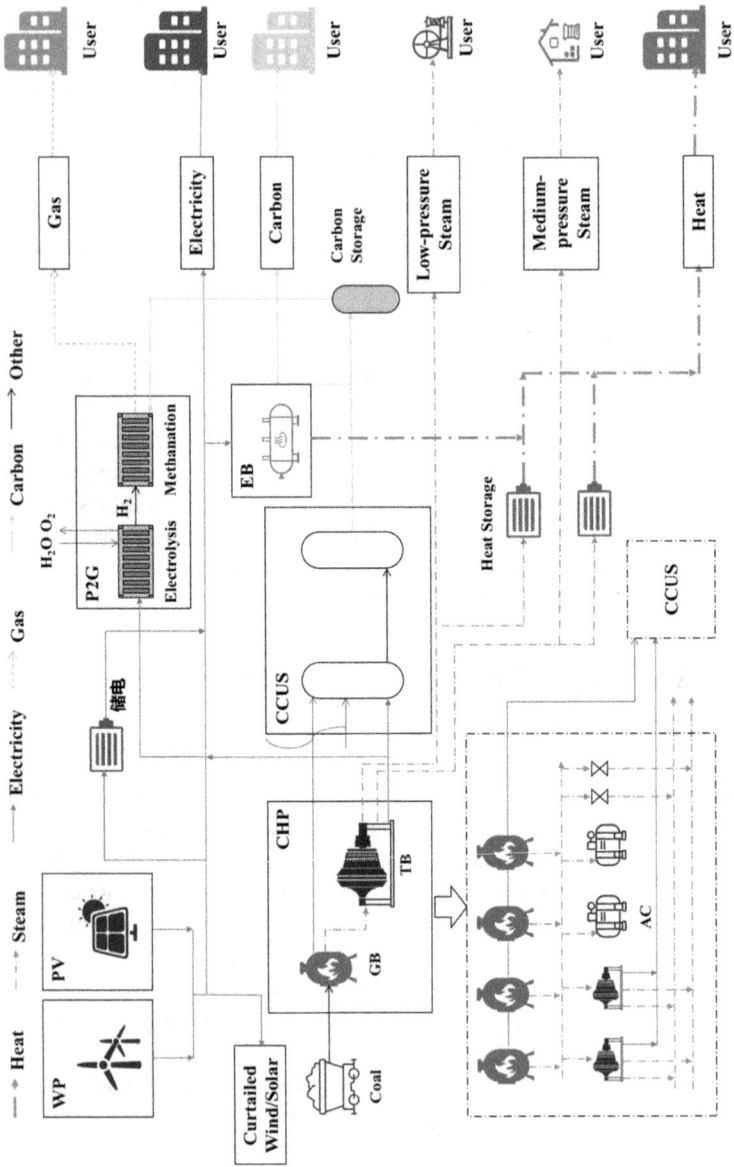

FIGURE 3.5   The structure of the integrated energy supply system studied in this chapter.

TABLE 3.1   Energy demand settings for the next hour.

PARAMETER	UNIT	VALUES				ACHIEVEMENTS STATUS
		T_1	T_2	T_3	T_4	
$v_{supply,0}$	t/h	295.00	295.00	295.00	295.00	√
$v_{supply,1}$	t/h	75.00	75.00	75.00	75.00	√
$v_{supply,2}$	MW	0.00	0.00	0.00	0.00	×
$v_{supply,3}$	t/h	0.00	0.00	0.00	0.00	×
$v_{supply,4}$	MW	60.00	60.00	60.00	60.00	√
$v_{supply,5}$	t/h	0.00	0.00	0.00	0.00	×

TABLE 3.2   Settings of equipment lag parameters.

PARAMETERS	DURATION (MIN)
τ_1	5
τ_2	15
τ_3	4
τ_4	2

TABLE 3.3   Upper and lower limit settings of system parameters.

PARAMETER	LOWER LIMIT	UPPER LIMIT	PARAMETER	LOWER LIMIT	UPPER LIMIT
v_{coal} (t/h)	0	200	$P_{sx,0}$ (t/h)	0	10
vo_{CHP1} (%)	0	100	$P_{sx,1}$ (MW)	0	10
vo_{CHP2} (%)	0	100	$P_{sx,2}$ (MW)	0	10
$v_{ccus,in2}$ (MW)	0	5	$P_{sx,3}$ (MW)	0	10
$v_{ccus,out0}$ (t/h)	0	10	$P_{sx,4}$ (MW)	0	10
$v_{P2G,in0}$ (t/h)	0	10	$P_{sx,5}$ (MW)	0	10
$v_{P2G,in1}$ (MW)	0	5	P_{CHP} (MW)	0	70
$P_{EB,in0}$ (MW)	0	5	P_{P2G} (t/h)	0	5
$v_{TP,out1}$ (t/h)	0	480	P_{EB} (MW)	0	5
$v_{TP,out0}$ (t/h)	0	160	—	—	—

TABLE 3.4　Energy supply and abandoned wind and solar power values at each demand time point after optimized calculation.

PARAMETER	T_1	T_2	T_3	T_4
$v_{supply,0}$ (t/h)	295.00	295.00	295.00	295.00
$v_{supply,1}$ (t/h)	75.00	75.00	75.00	75.00
$v_{supply,2}$ (MW)	0.00	0.00	5.11	13.05
$v_{supply,3}$ (t/h)	9.53	13.92	13.10	17.24
$v_{supply,4}$ (MW)	60.00	60.00	60.00	60.00
$v_{supply,5}$ (t/h)	1.64	2.59	3.05	3.05
$P_{curtailment}$ (MW)	0.00	0.00	0.00	0.00

The operational stability at time t is defined as $I_{smo,t}$, and its calculation method is as follows:

$$I_{smo,t} = \left(1 - \frac{|u_{m,t} - u_{m,t-1}|}{u_i}\right) \times 100 \tag{3.36}$$

The average operational stability is hereby defined as follows, which represents the average value of operational stability at each moment.

$$\overline{I_{smo}} = \frac{1}{T} \sum_{t=1}^{T} I_{smo,t} \tag{3.37}$$

Following a thorough examination of the discrepancies in multi-energy properties, there was an observed enhancement in the mean stability of coal feed volume, which increased from 95.32% to 98.14%, representing a 2.96% rise.

Figure 3.6 compares the differences in results between the proposed method in this chapter and the method that does not consider the effect of time delay. When the time lag is taken into consideration, a total coal consumption decrease of 121.26 t/h and an increase in total profit of 1.21×105 yuan is observed in the last three cycles. This finding suggests that the consideration of temporal discrepancies among diverse energy sources has the potential to curtail periodic coal consumption to a certain extent. An examination of the broader optimization cycle reveals a 72.00 t/h escalation in coal consumption. This is due to the fact that, following consideration of the lag model, it is not possible to optimize the operation of the cogeneration unit in the previous time period.

Specifically, as demonstrated in Figure 3.7, due to the initial value limit on the low-pressure steam flow of the cogeneration unit in the previous period, in order to meet the low-pressure steam flow demand corresponding to the first

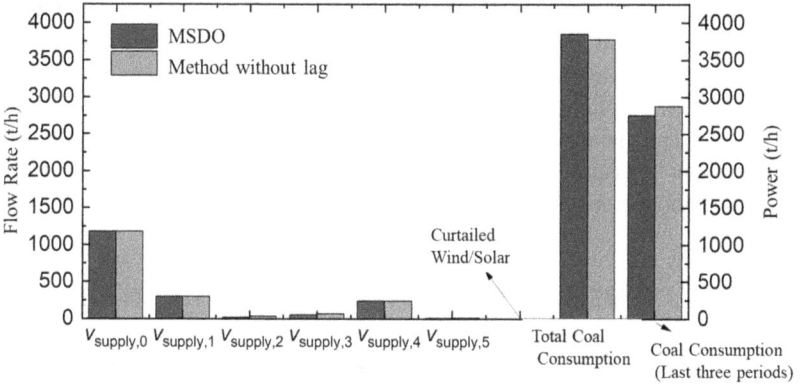

FIGURE 3.6 Comparison of optimization results without considering multifunctional lag.

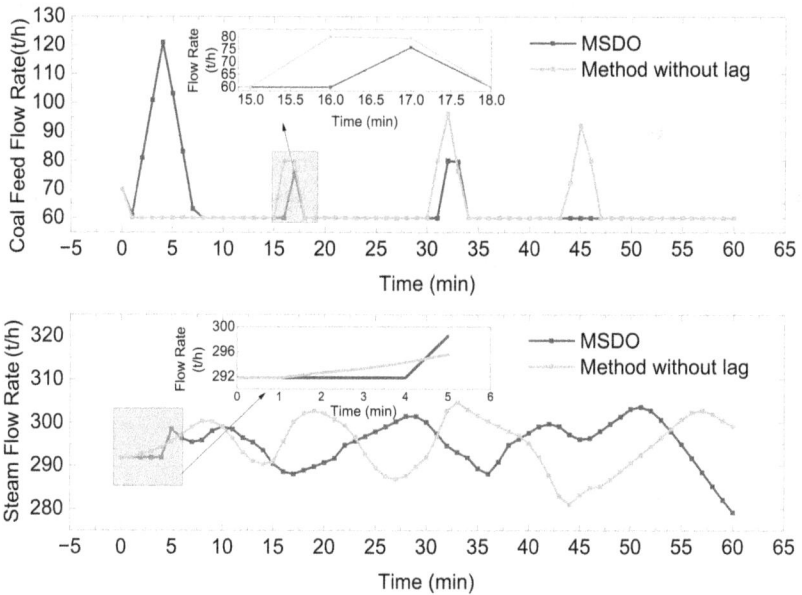

FIGURE 3.7 Comparison of coal consumption and low-pressure steam output before and after optimization.

calculation sub-cycle, the steam output after the period had to be increased, resulting in an increase in the average coal consumption at the beginning of this sub-cycle. It is evident that, commencing from the final three calculation sub-cycles, this method has the potential to markedly curtail the coal consumption of the system.

This diagram mainly describes the differences between the proposed MSDO method and the method that does not consider equipment lag performance in terms of load supply volume, wind and solar power curtailment rate, total coal consumption, etc.

This graph mainly describes the differences in the results of fuel feed flow rate and low-pressure steam flow rate between the MDSO method and the method without considering lag.

3.4.3 Analysis of lag duration and minimum time granularity

The elastic space was uniformly set to five unit values. As demonstrated in Figure 3.8, the impact of the response lag duration of different devices on the total profit increment is revealed. The findings indicate that for cogeneration units and carbon capture equipment, an augmentation in the time lag length facilitates an enhancement in the overall dispatching profit. Conversely, the electric-to-gas equipment and electric boilers demonstrate an inverse trend, wherein the reduced time lag is more conducive to enhancing profitability. With regard to the order of magnitude of overall profit changes, the time lag of cogeneration units has the most significant impact on the overall dispatching profit. It has been demonstrated that an increase of 1 min in time lag can result in an approximate increase of 250.53 yuan in the total profit of the subsequent three cycles. Conversely, the time lag inherent to electric boilers exerts

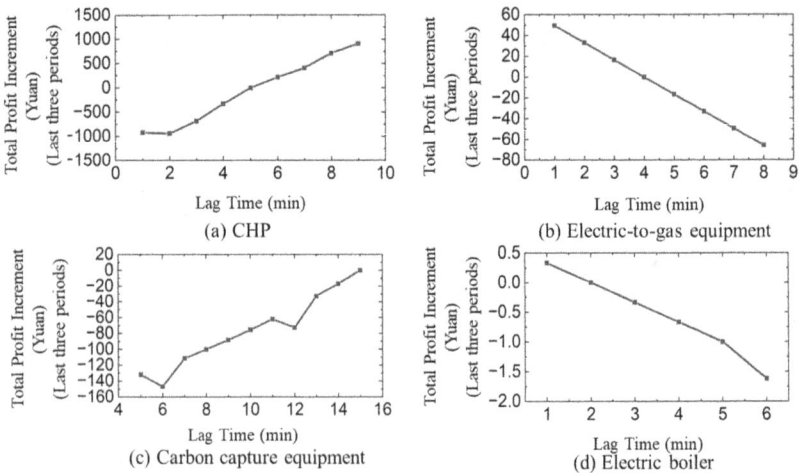

FIGURE 3.8   The impact of equipment lag duration on system scheduling profits.

a negligible influence on overall profitability. This finding suggests that the effective calculation of the time lag of cogeneration units facilitates the comprehension of the benefits of the overall dispatching plan and is conducive to the effective implementation of the dispatching plan.

This chart mainly describes the impact of different lag durations in cogeneration units, electric-to-gas equipment, carbon capture equipment, and electric boilers on the total profit increment.

The objective is to minimize the impact of the time granularity value on the computational efficiency and benefits of the scheduling scheme. In this context, computational efficiency is defined as the time required to calculate the scheduling scheme under the same computational conditions. That is to say, it is the computational time consumption.

As demonstrated in Figure 3.9, the investigation focuses on the impact of varying minimum time granularities on the calculation time consumption. It has been demonstrated that an augmentation in the minimum time granularity has the capacity to reduce the calculation time consumption, exhibiting a power function relationship.

This graph mainly describes the relationship between the minimum time granularity and the calculation time consumption. It can be found that after the minimum time granularity is less than or equal to 60 s, the calculation time consumption remains almost unchanged.

When the minimum time granularity is set to 1 s, the computing time reaches 102.65 s, which is approximately 218 times that of 60 s. However, when the minimum time granularity was set to 5 s, the computing time rapidly decreased to 9.73 s, which was only 20.67 times that of 60 s. When the minimum time granularity is greater than 60 s, the difference in calculation time is not significant.

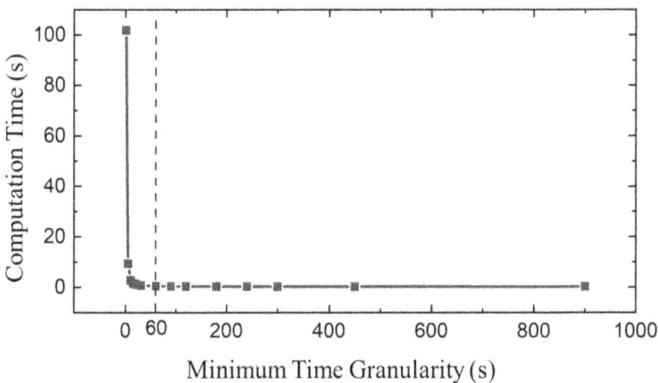

FIGURE 3.9   Granularity of minimum time and calculation time.

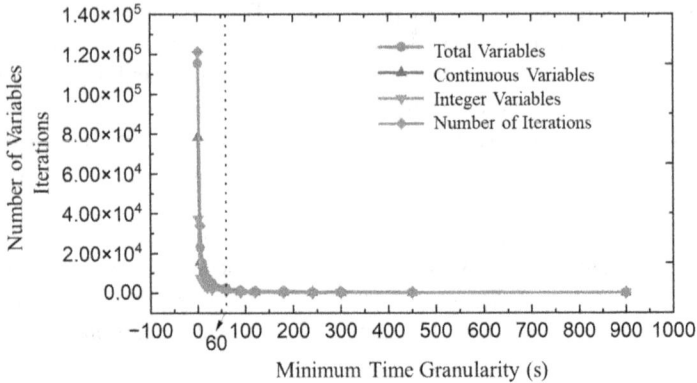

FIGURE 3.10 Minimum time granularity and model complexity.

As illustrated in Figure 3.10, there is a direct correlation between the minimum time granularity and the complexity of the model. In this model, the complexity is characterized by the variables and the number of iterations. It is evident that the total number of variables and the number of iterations in the model exhibit a power function relationship with the increase of the minimum time granularity. It is evident that as the minimum time granularity increases, the complexity of the model decreases. When the minimum time granularity is set to 1 s, the total number of variables reaches 1.15×10^5. Conversely, when the minimum time granularity is set to 60 s, the total number of variables is reduced to 2562.

This graph mainly describes the relationship between the minimum time granularity and the total number of variables, the number of continuous variables, the number of integer variables, and the number of iterations. It can be found that a minimum time granularity of 60 s is a turning point.

With regard to computational accuracy, the MSDO method proposed in this chapter can accurately predict the output of the device at different minimum time granularities. As illustrated in Figure 3.12, the low-pressure steam output of a CHP unit is used as an exemplar to demonstrate the impact of minimum time granularity on prediction accuracy. The prediction accuracy is characterized by the average prediction deviation. The actual value is derived through the optimization solution, while the predicted value is calculated according to steam flow conservation formula. The results are displayed in Figure 3.11. Despite the fact that the prediction error marginally increases when the minimum time granularity is elevated from 15 s to 4 min, the mean error obtained by employing the MSDO method proposed in this study remains less than 1%.

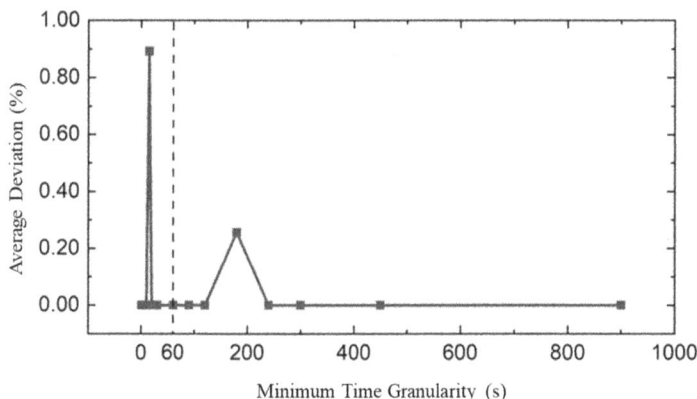

FIGURE 3.11   Minimum time granularity and model accuracy.

This graph mainly describes the relationship between the minimum time granularity and the average deviation.

As demonstrated in Figure 3.12, the impact of the minimum time granularity on the overall scheduling profit is evident. Specifically, Figure 3.12 (a) represents the total profit (i.e., the profits of the four periods), while Figure 3.12 (b) represents the profits of the last three periods. The rationale behind the separation of Figures 3.12 (a) and (b) is that following the introduction of the lag parameter, an initial value limit becomes apparent in the calculation of the first cycle. As demonstrated in Figures 3.12 (a) and (b), the lower half of each figure corresponds to an enlarged view of the selected area in the upper half. When the time granularity is greater than 60 s, the relationship between scheduling profit and time granularity can be observed more clearly.

This graph mainly describes the relationship between the minimum time granularity and the total profit increment, and makes a comparison from two scenarios: the full cycle and the first three cycles.

The results indicate that as the minimum time granularity decreases, the overall profit increases, and conversely, as the minimum time granularity increases, the overall profit also increases. For instance, an increase in the minimum time granularity from 1 s to 60 s, as implemented in this chapter, has the potential to enhance total profit by 1.81×105 yuan. This is primarily due to the fact that the reduction of minimum time granularity results in an increase in the total coal consumption within the cycle. It is well established that coal consumption exerts a significant impact on the profit of the entire dispatching cycle.

It has been demonstrated that as the minimum time granularity exceeds 60 s, the overall profit first increases and then decreases slightly. Within the time granularity range of 1 min to 3 min, an increase in time granularity has been

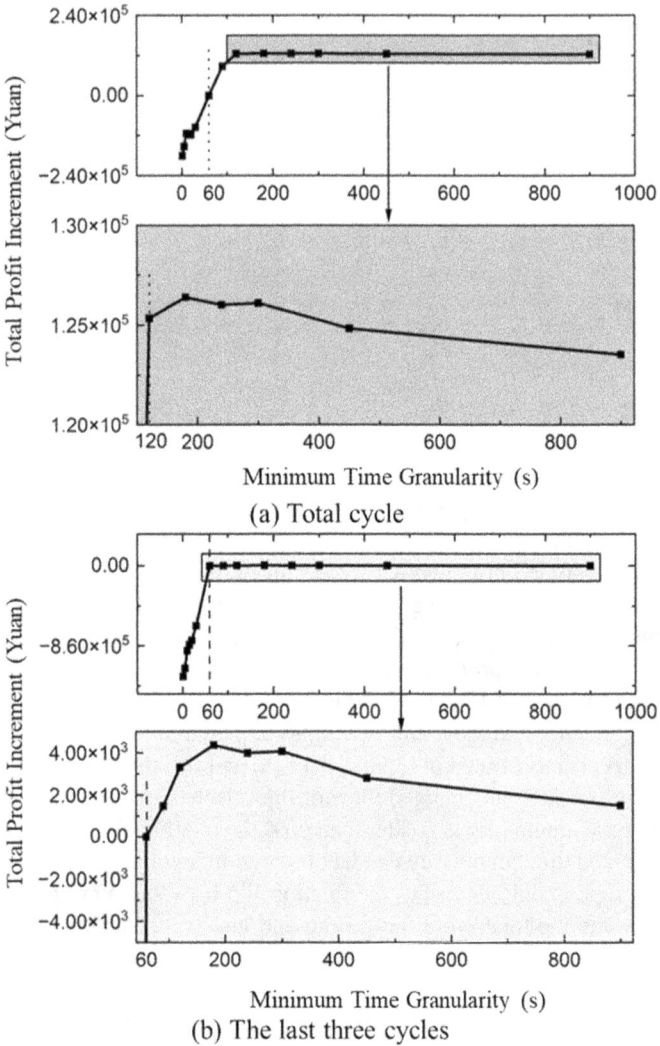

FIGURE 3.12   Minimum time granularity and profit increment.

shown to lead to an overall improvement in profit. Should the lag effect of the electric boiler be disregarded, there will be a marginal increase in overall profit. However, when the minimum time granularity was increased to 4 min, the overall profit declined compared with 5 min. This finding suggests that disregarding the lag effect of the electric-to-gas equipment is also conducive to enhancing overall profits, which is consistent with the previous conclusion.

It is evident that as the minimum time granularity exceeds 5 min, the overall profit undergoes a decline. This indicates that disregarding the lag effect of cogeneration units and carbon capture equipment will not result in an enhancement of the overall profit. It is evident that an increase in the time granularity leads to a reduction in model complexity and an enhancement in profits. However, it is imperative to consider the lag effect of specific energy sources to achieve optimal profit enhancement. The optimal minimum time granularity is at the minute level. For the lag parameters set out in this chapter, the optimal minimum time granularity is 3 min.

As illustrated in Figure 3.13, the minimum time granularity is shown to be related to varying demand cycle lengths, T, which are categorized into 1 h, 2 h, 5 h, and 10 h, respectively. As demonstrated in Figure 3.13, when the minimum time granularity is less than 60 s, the computing time increases in proportion to

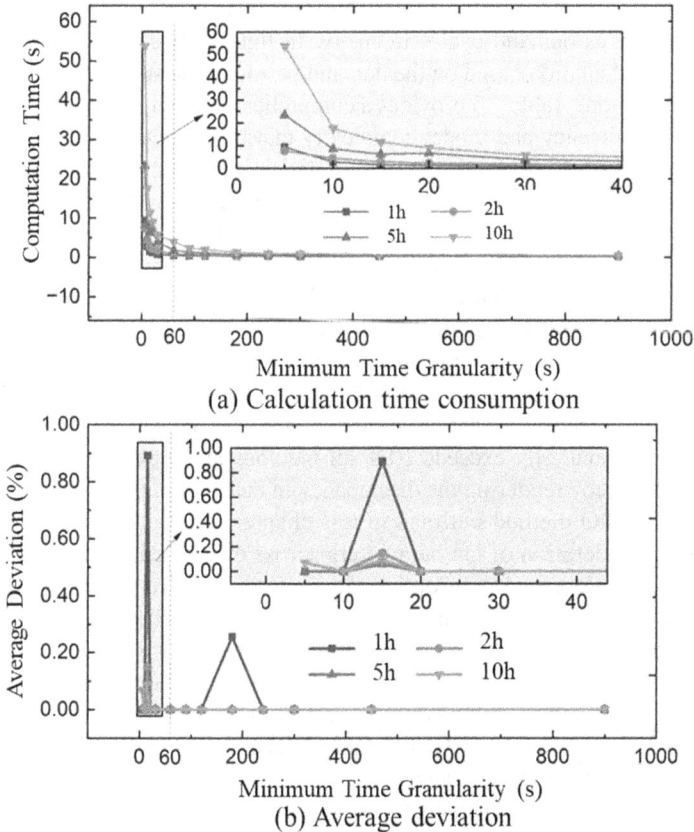

(a) Calculation time consumption

(b) Average deviation

FIGURE 3.13 Minimum time granularity and demand cycle length.

the demand cycle length. This effect is particularly pronounced when the time granularity is minimal.

This graph is divided into two subgraphs. Subgraph (a) mainly describes the relationship between the minimum time granularity and the calculation time, while subgraph (b) mainly describes the relationship between the minimum time granularity and the average deviation.

To illustrate this, consider a scenario in which the minimum time granularity is set to 5 s. In the event of an increase in the length of the required period from 1 h to 10 h, the calculation time will increase by a factor of 4.72. However, when the minimum time granularity is set to 1 s, the calculation of the 10-h demand period cannot be completed on the computer used in this chapter. From the perspective of average deviation, under the same time granularity, an increase in the length of the demand cycle will lead to a slight reduction in the average deviation, but this difference is not significant, all within 1%. It is acknowledged that the accuracy of long-term predictions may be subject to rapid decline as demand cycles intensify. In light of this, the present study employs a calculation based on the demand period of the subsequent hour.

Furthermore, Table 3.5 provides a comprehensive comparison of the computational efficiency and model complexity of various methods across different time granularities. As the time granularity increases, a consistent trend emerges across all methods: the greater the time granularity, the higher the computational efficiency and the lower the complexity of the model. When the minimum time granularity is set between 1 s and 5 s, Method #1 does not successfully calculate the mean, but the MSDO method and Method #0 do. Furthermore, the MSDO method involves fewer variables. From the perspective of overall model complexity, the complexity of the MSDO method is close to that of Method #0 and significantly higher than that of Method #1. However, this does not linearly increase the computing time. In scenarios where the minimum time granularity exceeds 10 s, all methods are capable of responding promptly, thereby rendering the discrepancy in calculation time negligible.

The MSDO method outlined in this chapter is notable for its comprehensive consideration of lag parameters across diverse energy sources. The multi-time scale scheduling optimization in the integrated energy supply system has been achieved through a single-layer model, which has advantages in terms of robustness, feasibility, and the capacity to consume renewable energy. Conversely, the user demand response and cumulative demand elasticity proposed in this chapter, as deduced from the demand elasticity analysis, have the potential to enhance the efficiency of the integrated energy supply system. The analysis of the minimum time granularity also indicates that when considering the lag characteristics of energy, it is necessary to reasonably select the minimum time granularity to balance benefits and feasibility.

TABLE 3.5 Comparison of computational efficiency and model complexity of various multi-time scale methods.

MINIMUM TIME GRANULARITY (S)	CALCULATION TIME CONSUMPTION (S)			TOTAL NUMBER OF VARIABLES		
	METHOD #0 (LAG NOT CONSIDERED)	METHOD #1 (DIRECTLY CONSIDER LAG)	MSDO METHOD	METHOD #0 (LAG NOT CONSIDERED)	METHOD #1 (DIRECTLY CONSIDER LAG)	MSDO METHOD
1	148.15	—	101.86	1.51×10^5	—	1.15×10^5
5	7.84	—	9.40	2.40×10^4	—	2.30×10^4
10	2.84	2.75	2.84	1.21×10^4	1.52×10^4	1.52×10^4
15	1.56	1.72	1.58	1.01×10^4	5399	1.01×10^4
20	1.15	1.37	1.28	7602	4101	7602
30	0.79	1.03	0.78	4016	2783	5082
60	0.54	0.65	0.48	2562	1433	2562
90	0.45	0.50	0.40	1355	935	1312
120	0.34	0.45	0.36	946	670	913
180	0.32	0.36	0.35	882	445	882
240	0.28	0.31	0.30	343	268	389
300	0.28	0.33	0.29	546	546	546
450	0.33	0.27	0.27	250	51	245
900	0.23	0.26	0.23	210	19	113

IES Integration of Carnot Battery

4

Xiaojie Lin, Ziying Zhao, Haonan Zheng, Peng Sun, and Jiahao Xu

4.1 INTEGRATION OF CARNOT BATTERIES WITH INTEGRATED ENERGY SYSTEM (IES)

4.1.1 The structure of industrial IES

The global industrial sector stands at a critical juncture, compelled by a confluence of economic pressures and environmental imperatives to fundamentally rethink its relationship with energy. The long-standing paradigm of inexpensive and abundant fossil fuels is being challenged by volatile energy markets and an increasingly urgent mandate to decarbonize operations. This situation presents a complex conundrum for industrial operators, who must simultaneously manage production costs, ensure energy supply reliability, and meet stringent environmental targets. In response to this multifaceted challenge, the concept of the industrial IES has moved from a theoretical framework to a practical necessity. It represents a paradigm shift away from viewing energy inputs as isolated commodities and toward a holistic approach that optimizes the intricate web of energy conversion, storage, and utilization across an entire facility. Within the context of a manufacturing plant, a chemical processing facility,

DOI: 10.1201/9781003630821-4

or a large-scale food production line, energy is the lifeblood of production. A constant and reliable supply of electricity is essential to power mechanical equipment, intricate control systems, and facility lighting. Concurrently, thermal energy, often delivered as hot water or high-pressure steam, is indispensable for a vast array of processes, including process heating, drying, chemical reactions, and sterilization.

However, a defining and often problematic characteristic of these environments is the generation of vast quantities of low-grade waste heat, typically from wastewater or exhaust gases at temperatures around 100°C. Conventionally, this thermal energy is difficult to recover due to its low temperature and density, leading to significant energy wastage and exergy destruction. Simultaneously, the push toward sustainability has led to an increased penetration of renewable energy sources like solar photovoltaic (PV) and wind power. While clean, their intermittent and unpredictable nature creates volatility that clashes with the typically stable and relentless energy demand of industrial processes. This mismatch necessitates intelligent energy dispatching and, critically, robust energy storage solutions. While electrochemical batteries are adept at managing short-term electrical fluctuations, they fail to address the crucial thermal dynamics of industrial systems. This creates a technological gap, highlighting the need for a solution that can holistically bridge the electrical and thermal domains. The Carnot battery emerges as a uniquely suitable technology to fill this void, offering a mechanism for bidirectional energy conversion that can simultaneously absorb surplus renewable electricity, recover and upgrade waste heat, and dispatch power or heat on demand. It is not merely a storage device but a synergistic enabler of a more flexible, efficient, and circular industrial energy economy.

4.1.2 The role of Carnot batteries in industrial IES

An industrial IES is a carefully designed and highly coordinated framework intended to optimize how different energy carriers are converted and utilized. Its purpose is to boost overall energy efficiency, reduce resource consumption, and minimize environmental impact. In a typical industrial setting, electricity and hot water serve as the two main energy carriers. Electricity powers machines, lighting, and motion systems, while hot water is indispensable for space heating, heat exchange, and specific industrial processes—particularly in sectors such as chemical engineering, metallurgy, and food processing.

A notable feature of many industrial operations is the large amount of low-grade waste heat they generate, often in the form of wastewater around 100°C. This heat originates from various sources, including exhaust gases, effluent streams, and equipment cooling losses. Because of its relatively low temperature and energy density, traditional recovery technologies struggle to make efficient use of it. However, recent advances in integrated and multi-energy complementary technologies have enabled IESs to capture and reuse this low-grade heat. Through high-efficiency combined heat and power (CHP) units, waste heat recovery systems, and heat pumps, what was once discarded can now be converted into useful heat or even electricity, significantly improving overall system performance.

A typical industrial IES brings together a diverse portfolio of generation, conversion, and storage technologies. Core components often include renewable sources such as PV panels and wind turbines (WTs), which harness local solar and wind resources to deliver clean power. Conventional units—like CHP plants and gas boilers (GBs)—provide stable supplies of both electricity and heat, usually fueled by natural gas. Electric boilers (EBs) introduce direct electro-thermal coupling, turning electricity straight into heat. To handle the intermittent nature of renewables and the fluctuating energy demand, storage systems are essential. While electrochemical batteries are effective for short-term power storage, they cannot adequately address the thermal dynamics and waste-heat utilization challenges inherent in industrial processes.

This technological gap highlights the urgent need for advanced energy storage solutions capable of bridging the electrical and thermal domains. Within this context, the Carnot battery has emerged as a promising option. It uniquely uses surplus electricity to upgrade low-grade waste heat through a heat pump mechanism, storing it as high-temperature thermal energy. The stored heat can later be flexibly deployed—either converted back into electricity through a thermodynamic cycle such as the Organic Rankine Cycle (ORC) or supplied directly as process heat. Therefore, integrating a Carnot battery is not merely adding another storage unit—it represents a strategic

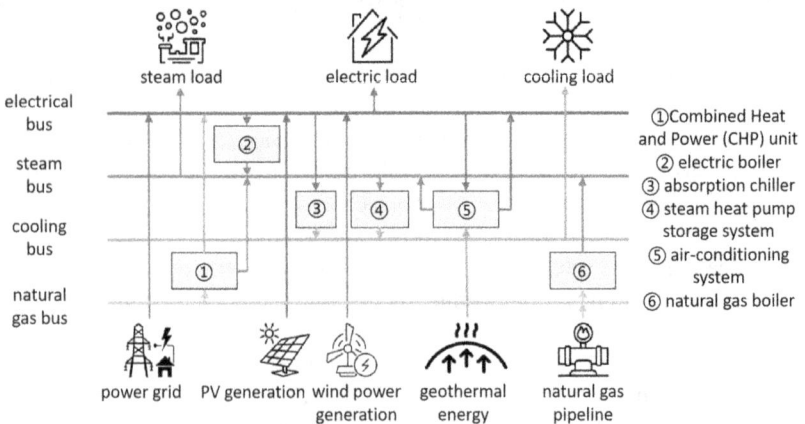

FIGURE 4.1 Structure of the industrial IES integrated with a Carnot battery.

enhancement that introduces new flexibility and synergy to the entire system. It enables greater absorption of surplus renewable power, efficient recovery and upgrading of valuable waste heat, and coordinated delivery of both electricity and thermal energy. In doing so, it tackles the core challenges of electro-thermal coupling and resource efficiency in modern industrial energy systems. Figure 4.1 illustrates the structure of a typical IES integrated with a Carnot battery.

4.2 MATHEMATICAL MODELING OF SYSTEM COMPONENTS

To analyze and optimize the complex energy flows within the IES, a detailed mathematical model for each component is essential. This section outlines the governing equations that describe the operational characteristics of the key energy generation, conversion, and storage units.

4.2.1 The Carnot battery

At a conceptual level, the Carnot battery functions like any other energy storage device: it absorbs electrical energy when it is plentiful and dispatches it when needed. However, this simple functional description belies the sophisticated process by which it operates. Unlike conventional electrochemical batteries that store energy directly in chemical bonds, the Carnot battery is a thermo-mechanical system that transforms electricity into thermal potential. Therefore, a deeper understanding of its value and operational characteristics requires an examination of its underlying thermodynamic principles, as it leverages distinct thermodynamic cycles to charge and discharge.

The charging process is fundamentally a high-temperature heat pump cycle. It consumes electricity—ideally low-cost off-peak grid power or surplus renewable energy—to drive a compressor, which elevates the pressure and temperature of a working fluid. This high-temperature fluid then transfers heat to the Thermal Energy Storage (TES) system. The working fluid, having released its heat, is expanded and passed through an evaporator, where it absorbs low-grade heat from an external source, such as industrial wastewater. This dual function of absorbing excess electricity while simultaneously recovering and upgrading waste heat constitutes its key strategic advantage. The performance of this cycle is quantified by the Coefficient of Performance (COP), which is the ratio of heat delivered to the TES to the electrical work consumed by the compressor. The magnitude of the "temperature lift"—the difference between the heat source temperature and the final storage temperature—is a primary determinant of the COP; a smaller lift generally results in a higher COP and more efficient charging.

Conversely, the discharging process operates as a heat engine, or power cycle, converting the stored thermal energy back into electricity. The high-temperature TES serves as the heat source for a thermodynamic cycle, commonly an ORC due to its effectiveness in converting medium-to-high temperature heat into work. The use of an ORC is particularly advantageous in this context because organic working fluids have lower boiling points and different thermodynamic properties than water, allowing for efficient power generation from heat sources that might be insufficient for a traditional steam Rankine cycle. In the ORC, the organic fluid is vaporized by the heat from the TES, and this high-pressure vapor drives a turbine or expander to generate mechanical work, which in turn drives a generator to produce electricity. The efficiency of this power cycle is a critical determinant of the overall system performance. This dual-output capability is a core feature, as the stored heat can also be extracted directly via a heat exchanger to provide high-quality process heat to the industrial facility.

The state of thermal storage is governed by the principle of energy conservation. The equation tracks the stored energy from one time step to the next, accounting for all energy flows into and out of the storage system.

$$E_{\text{sto}}(t+1) = E_{\text{sto}}(t) + \frac{1}{COP-1} P_{\text{hp}}(t) \Delta t - \frac{P_{\text{orc}}(t)}{\eta_{\text{orc}}} \Delta t - \frac{P_{\text{hex}}(t)}{\eta_{\text{hex}}} \Delta t \qquad (4.1)$$
$$- \lambda E_{\text{sto}}(t) \Delta t$$

$$E_{sto}^{L} \le E_{\text{sto}}(t) \le E_{sto}^{U} \qquad (4.2)$$

where E_{sto} is the state of thermal storage of the Carnot battery at time t (kWh), $P_{\text{hp}}(t)$ is the power consumed by the heat pump for thermal storage at time t (kW), $P_{\text{orc}}(t)$ is the power generated by the ORC during heat release at time t (kW). $P_{\text{hex}}(t)$ is the heat exchange power when the Carnot battery supplies heat directly at time t (kW). COP is the coefficient of performance of the heat pump, η_{orc} is the electrical efficiency of the ORC, η_{hex} is the efficiency of the direct heat exchanger. λ is the heat loss coefficient of the storage tank (%/h), Δt is the time step (h), E_{sto}^{L} and E_{sto}^{U} are the lower and upper limits of the thermal storage capacity (kWh), respectively. Notably, both the COP and η_{orc} change with operating conditions rather than remaining constant. Their values are derived from a multi-condition performance database that captures the dynamic behavior of system components under varying thermodynamic states.

The overall efficacy of a Carnot battery is typically assessed by its Round-Trip Efficiency (RTE), defined as the ratio of the total electricity discharged to the total electricity consumed. However, this single metric can be misleading. As a composite value, the RTE is influenced by the COP, the ORC efficiency, and thermal losses. While the electrical RTE of current Carnot battery designs is often lower than that of lithium-ion batteries, a direct comparison is incomplete. The Carnot battery's true value proposition extends beyond simple electrical storage. By providing valuable process heat as an alternative output and by upgrading waste heat (which has its own exergetic value), the Carnot battery can offer superior system-level economic and environmental benefits.

Furthermore, the selection of materials is critical. The choice of working fluids for the heat pump and ORC cycles must balance thermodynamic performance with factors like thermal stability, cost, and environmental impact. Similarly, the medium for the TES system—be it sensible heat storage in molten salts or solid-state materials like concrete, or latent heat storage in phase-change materials—dictates the operating temperature range, energy density, cost, and lifespan of the storage component. These material science considerations represent a key area of ongoing research and development.

4.2.2 Mathematical models of other IES components

For a thorough analysis of the integrated system, it is necessary to develop mathematical models for each major energy conversion and generation component in the IES.

4.2.2.1 PV system

The output power of a PV system mainly depends on its installed capacity, the level of solar radiation, and the surrounding temperature. Since it is a non-controllable source, its generation is usually treated as a fixed time-series input in energy hub modeling. The model is described as follows:

$$P_{pv}(t) = Pe_{pv} \cdot \eta_{pv}(t) \cdot \frac{G_t}{G_{ref}} \cdot \left[1 + K_T \left(T(t) - T_{ref} \right) \right] \tag{4.3}$$

where $P_{pv}(t)$ is the real-time power output of the PV system (kW), Pe_{pv} is the rated power of the PV system (kW), η_{pv} is the performance coefficient of the PV system, G_t is the hourly average solar irradiance (W/m²), G_{ref} is the solar irradiance under standard conditions (W/m²), K_T is the power temperature coefficient of the PV system, $T(t)$ is the operating temperature of the PV panel (°C), and T_{ref} is the reference temperature under standard conditions (°C).

4.2.2.2 WT system

Similar to the PV system, the power output of a WT is dependent on its rated capacity and the prevailing wind speed. A piecewise cubic polynomial function is commonly used to model its power curve. Its stochastic nature means it is also treated as a fixed time-series input. The model is given by:

$$P_{wt}(t) = \begin{cases} 0, & V(t) < V_{cut-in}, V(t) \geq V_{cut-out} \\ \dfrac{V_t^3 - V_{cut-in}^3}{V_r^3 - V_{cut-in}^3} \cdot Pe_{wt}, & V_{cut-in} \leq V(t) < V_r \\ Pe_{wt}, & V_r \leq V(t) < V_{cut-out} \end{cases} \tag{4.4}$$

where $P_{wt}(t)$ is the real-time power output of the WT (kW); Pe_{wt} is the rated power of the WT (kW); V_t is the hourly average wind speed (m/s), and V_{cut-in}, $V_{cut-out}$, V_r are the cut-in, cut-out, and rated wind speeds of the turbine (m/s), respectively.

4.2.2.3 CHP unit

The CHP unit is a critical component that simultaneously produces electricity and useful heat, significantly improving overall energy efficiency. The energy balance equations for a CHP unit are presented as follows:

$$Q_{CHP,fuel}(t) = F_{gas}(t) \cdot LHV_{gas} \tag{4.5}$$

$$P_{CHP,elec}(t) = \eta_{elec} \cdot Q_{CHP,fuel}(t) \tag{4.6}$$

$$Q_{CHP,heat}(t) = \eta_{heat} \cdot Q_{CHP,fuel}(t) \tag{4.7}$$

where $Q_{CHP,fuel}(t)$ is the instantaneous fuel energy input (MJ); $F_{gas}(t)$ is the instantaneous natural gas volume flow rate (m³/s); LHV_{gas} is the lower heating value of natural gas (MJ/m³); $P_{CHP,elec}(t)$ and $Q_{CHP,heat}(t)$ are the electrical and thermal power outputs, respectively (kW); η_{elec} is the electrical efficiency; and η_{heat} is the thermal efficiency.

The operation of the CHP unit is subject to several constraints:

$$P_{CHP,elec}^{min} \leq P_{CHP,elec} \leq P_{CHP,elec}^{max} \tag{4.8}$$

$$\alpha(t) = \frac{P_{\text{CHP,elec}}(t)}{Q_{\text{CHP,heat}}(t)} \in \alpha_{\min}, \alpha_{\max} \tag{4.9}$$

where $P_{\text{CHP,elec}}^{\min}$ and $P_{\text{CHP,elec}}^{\max}$ are the minimum and maximum electrical power output, respectively (kW), and $\alpha(t)$ is the heat-to-power ratio at time t.

4.2.2.4 Gas boiler

The GB serves as a flexible heat source capable of efficiently transforming the fuel's chemical energy into usable thermal energy. It mainly functions as a backup during CHP system downtime and as a supplementary unit for handling unexpected surges in heat demand. The governing equations are presented as follows:

$$Q_{\text{GB,fuel}}(t) = F_{\text{gas,GB}}(t) \cdot \text{LHV}_{\text{gas}} \tag{4.10}$$

$$Q_{\text{GB,out}}(t) = \eta_{\text{GB}} \cdot Q_{\text{GB,fuel}}(t) \tag{4.11}$$

where $Q_{\text{GB,fuel}}(t)$ and $F_{\text{gas,GB}}(t)$ refer to the energy supplied by the fuel and the corresponding natural gas flow rate. $Q_{\text{GB,out}}(t)$ is the boiler's thermal output (kW), while η_{GB} stands for the thermal efficiency.

The boiler's operation is subject to its own performance and capacity constraints:

$$Q_{\text{GB}}^{\min} \le Q_{\text{GB,out}}(t) \le Q_{\text{GB}}^{\max} \tag{4.12}$$

4.2.2.5 Electric boiler

As an essential electro-thermal coupling device, the EB converts electricity into thermal energy. During periods when the CHP unit runs in a power-dominant mode, it can utilize excess renewable power while simultaneously meeting additional heating requirements. Its energy conversion process can be described as follows:

$$P_{\text{EB,in}}(t) = \frac{Q_{\text{EB,out}}(t)}{\eta_{\text{EB}}} \tag{4.13}$$

where $P_{\text{EB,in}}(t)$ is the electrical power input (kW), $Q_{\text{EB,out}}(t)$ is the thermal output (kW), and η_{EB} is the electric-to-thermal conversion efficiency.

4.2.2.6 Battery energy storage system (BESS)

The BESS mainly serves to improve the quality and stability of electrical energy within the subsystem. Thanks to its rapid response, it is well-suited for handling short-term fluctuations, though its upfront cost is comparatively high. The charging and discharging power are subject to the following constraints:

$$0 \le P_{es}^{ch}(t), P_{es}^{dis}(t) \le Pe_{es} \tag{4.14}$$

$$SOC_{es}(t) = SOC_{es}(t-1) \cdot (1 - \sigma_{es}) + \left(P_{es}^{ch}(t)\eta_{es}^{ch}(t) - \frac{P_{es}^{dis}(t)}{\eta_{es}^{dis}(t)} \right) \cdot \frac{\Delta t}{W_{es}} \tag{4.15}$$

$$SOC_{es}^{L} \le SOC_{es}(t) \le SOC_{es}^{U} \tag{4.16}$$

$$W_{es} = \rho_{es}Pe_{es} \tag{4.17}$$

where $P_{es}^{ch}(t)$ and $P_{es}^{dis}(t)$ are the charging and discharging power, respectively (kW); Pe_{es} is the rated power of the battery (kW); $SOC_{es}(t)$ is the State of Charge at time t; σ_{es} is the self-discharge coefficient; Δt is the time step (h); W_{es} is the rated storage capacity (kWh); η_{es}^{ch} and η_{es}^{dis} are the charging and discharging efficiencies, respectively; SOC_{es}^{L} and SOC_{es}^{U} are the lower and upper bounds of the SOC; and ρ_{es} is the power-to-capacity conversion coefficient (kW/kWh).

Building detailed mathematical models for each component lays a solid foundation for evaluating and optimizing the overall performance of the IES. Through this integrated framework, the intricate energy flows and interactions introduced by the integration of the Carnot battery can be accurately examined.

4.3 SYSTEM PERFORMANCE EVALUATION AND OPTIMIZATION METHODOLOGY

To make the most of integrating a Carnot battery within an industrial IES, a well-structured framework for performance assessment and optimization is urgently needed. The IES is inherently complex, characterized by multiple interconnected components and tightly coupled energy flows, which calls for a detailed approach to identify the best capacity configuration of the system's assets. The design and sizing of the Carnot battery, in particular, are influenced by several factors—including the quality of available waste heat and the extent of imbalance between energy supply and demand. Hence, a systematic optimization strategy is required to coordinate both system planning and operation, enabling the Carnot battery to work efficiently alongside other energy conversion units while achieving overall economic and environmental goals.

This section introduces a two-stage optimization approach for capacity configuration, developed to tackle this challenge. The method decouples long-term investment planning from short-term operational scheduling, creating a hierarchical framework that ensures both computational efficiency and realistic decision-making. Furthermore, a set of economic and environmental metrics is established to quantitatively assess the performance of various system configurations.

4.3.1 Two-stage capacity configuration optimization method

A two-level optimization framework is developed to tackle the complex planning challenges of an IES incorporating a Carnot battery. The method captures the strong coupling between long-term investment decisions—such as capacity planning—and short-term operational scheduling. The upper layer focuses on strategic planning, targeting the simultaneous minimization of annualized total cost (ATC) and carbon emissions. Meanwhile, the lower layer deals with day-to-day operation scheduling, aiming to reduce the annual operational cost based on the capacity configuration obtained from the upper layer.

The hierarchical structure can be expressed in mathematical form as follows: At the upper level, the goal is to find an optimal decision vector x, which denotes the capacity allocation of different devices, to minimize the objective function $F(x, y_1, y_2, \ldots, y_m)$ under the planning constraints $G(x) \leq 0$. At the lower level, given x, the model determines the optimal operational vector y, representing the hourly dispatch plan of each unit, to minimize its own objective function $F(x, y_1, y_2, \ldots, y_m)$ while satisfying the operational constraints $g(x, y_1, y_2, \ldots, y_m) \leq 0$.

$$\begin{cases} J_1 = \min_{x} F(x, y_1, y_2, \ldots, y_m) \\ s.t. \ G(x) \leq 0 \end{cases} \tag{4.18}$$

$$\begin{cases} J_2 = \min_{y} F(x, y_1, y_2, \ldots, y_m) \\ s.t. \ g(x, y_1, y_2, \ldots, y_m) \leq 0 \end{cases} \tag{4.19}$$

The optimization process starts by gathering input data such as the technical and economic parameters of candidate technologies, load demand curves, weather information, and energy price profiles. These parameters are then assigned to different layers of the model: equipment characteristics are fed into the upper layer, while energy data inform the operation of the lower layer. By iteratively solving and coordinating the two levels, the framework produces a series of Pareto-optimal capacity configurations, each reflecting a unique balance between economic performance and environmental impact.

4.3.1.1 Upper-level planning model

The upper-level model handles the strategic planning of installed capacities for all components in the IES. It evaluates economic feasibility using an overall cost assessment and incorporates a carbon emission model to quantify the environmental effects. Based on these considerations, a multi-objective optimization problem is established, as shown below:

$$Obj_{upper,f_1(x)} = \min\left(IC + OC + MC\right) \tag{4.20}$$

$$Obj_{upper,f_2(x)} = \min\left(ECM\right) \tag{4.21}$$

where $Obj_{upper,f_1(x)}$ and $Obj_{upper,f_2(x)}$ are the economic and environmental objective functions for the upper-level model, respectively. IC stands for the annualized investment cost (RMB/year), OC is the annual operating cost (RMB/year) obtained from the lower-level model, MC denotes the annual maintenance cost (RMB/year), and ECM represents the total amount of carbon emissions generated annually (tons of CO_2).

The Annualized Investment Cost (IC) is calculated by amortizing the initial capital cost of each piece of equipment over its operational lifetime using a capital recovery factor:

$$IC = \sum_{i \in Eq} Co_i \cdot Pe_i \cdot \frac{IR\left(1 + IR\right)^{l_i}}{\left(1 + IR\right)^{l_i} - 1} \tag{4.22}$$

where Co_i is the unit initial investment cost of equipment i (RMB/kW), Pe_i is the rated capacity of equipment i (kW), IR is the discount rate (%), l_i is the operational lifetime of equipment i (years), and Eq is the set of all equipment types in the system.

The Annual Maintenance Cost (MC) is typically estimated as a fixed percentage of the initial investment cost, annualized in the same manner:

$$MC = \sum_{i \in E} \zeta_i \cdot Co_i \cdot Pe_i \cdot \frac{IR\left(1 + IR\right)^{l_i}}{\left(1 + IR\right)^{l_i} - 1} \tag{4.23}$$

where ζ_i is the maintenance coefficient for equipment i.

The Annual Carbon Emissions (ECM) are attributed to the consumption of natural gas and the purchase of electricity from the grid, calculated as follows:

$$ECM = \sum_{i \in E} \varepsilon \cdot \left(\theta \cdot Q_{fuel} + \beta \cdot P_{ele} \right) \tag{4.24}$$

where ε is the carbon emission coefficient per unit of fuel consumed (tCO_2/kWh), Q_{fuel} is the total fuel consumed at hour t (kWh), β is the equivalent carbon emission coefficient for purchased electricity (tCO_2/kWh), and P_{ele} is the electricity purchased (kW).

4.3.1.2 Lower-level scheduling model

In the two-stage optimization planning model, the lower-level operational scheduling model conducts an hourly optimization of the IES, using the energy hub model as a constraint. Its objective function is the Annual Operating Cost (OC), which includes the cost of purchasing energy from external energy networks and the subsidies for selling energy back to those networks.

$$Obj_{lower, f_1(x)} = \min \left(OC \right) \tag{4.25}$$

The model is subject to constraints for each unit, including rated power limits.

4.3.2 Solution algorithm: Non-dominated Sorting Genetic Algorithm II (NSGA-II)

To solve the multi-objective optimization problem presented in the upper-level planning model, which involves the conflicting objectives of economic and environmental performance, it is necessary to employ an algorithm capable of effectively navigating such trade-offs. The solution to this type of problem is not a single optimal point, but rather a set of solutions known as the Pareto-optimal front. Within this set, any improvement in one objective can only be achieved at the expense of the other. To precisely identify this frontier of optimal trade-offs, this study utilizes the NSGA-II. This algorithm is widely recognized and applied in the field of engineering optimization for its excellent global search capabilities, its efficient mechanism for maintaining solution diversity, and its robust elite-retention strategy.

The operational flow of NSGA-II simulates the principles of natural selection, iteratively refining a population of candidate solutions. Its core mechanism can be decomposed into several key steps:

The process commences with Initialization. The algorithm randomly generates an initial population, wherein each "individual" represents a complete capacity configuration for the industrial IES. Concurrently, key algorithmic parameters are defined, including the population size, the maximum number of generations, and the probabilities of crossover and mutation. These parameters collectively govern the algorithm's search breadth and convergence speed.

This is followed by Fitness Evaluation. For each individual in the population (i.e., each capacity configuration), the lower-level scheduling model is invoked to perform a full 8760-h annual operational simulation. Upon completion, the corresponding OC and ECM are calculated. These two values serve as the core metrics for assessing the "fitness" of the individual.

The next stage involves the algorithm's core Elite Selection Strategy, which is composed of two parts: Non-dominated Sorting and Crowding Distance Calculation. Non-dominated sorting categorizes the entire population into different fronts based on Pareto dominance. Solutions located on the first front constitute the set of non-dominated solutions within the current population; they are not inferior to any other solution with respect to all objectives. To preserve diversity and prevent premature convergence to a localized region of the Pareto front, the algorithm calculates the crowding distance for each solution. This metric quantifies the density of surrounding solutions, and the algorithm gives preference to solutions in sparser regions, thereby ensuring that the final Pareto front is uniformly and broadly distributed across the objective space.

The final step is Genetic Operation. The algorithm generates the next generation of the population through three primary operations: selection, crossover, and mutation. The selection operator prioritizes individuals for reproduction based on their non-domination rank and crowding distance. Crossover simulates biological reproduction by combining the "genes" (i.e., equipment capacity parameters) of two parent solutions to create new offspring. Mutation introduces random alterations to an individual's parameters at a certain probability, endowing the algorithm with the ability to escape local optima and explore new regions of the solution space.

This entire process is repeated iteratively until the predefined maximum number of generations is reached. The final output of the algorithm is the set of all non-dominated solutions identified throughout the process, which collectively form the Pareto-optimal front for the optimization problem.

4.3.3 System performance evaluation metrics

To conduct a quantitative and multi-dimensional assessment of the optimization results, a clear and comprehensive framework of performance indicators is required. This framework is structured around the two core dimensions of economic and environmental performance, providing a standardized basis for the subsequent comparative analysis of scenarios and sensitivity analysis.

In the economic domain, the primary metric is the ATC. This is a comprehensive financial indicator that reflects the total economic burden of the system over its lifecycle. It is calculated as follows:

$$ATC = IC + MC + OC \tag{4.26}$$

The calculation for the Annualized IC and Annualized MC has been previously defined. To more intuitively measure the economic benefit of a given configuration relative to a baseline case (e.g., a system without energy storage), the Cost Saving Rate (CSR) is introduced:

$$CSR = \frac{ATC_{base} - ATC_{scenario}}{ATC_{base}} \times 100\% \tag{4.27}$$

where ATC_{base} is the annualized total cost of the baseline scenario and $ATC_{scenario}$ is the cost of the scenario being evaluated.

In the environmental domain, the core metric is the ECM. This indicator quantifies the direct impact of the system's operation on climate change, encompassing both direct emissions from the combustion of natural gas and indirect emissions attributed to the purchase of electricity from the grid:

$$ECM = \sum_{t=1}^{8760} \left(\theta \cdot Q_{fuel}^t + \beta \cdot P_{elec}^t \right) \tag{4.28}$$

where θ and β are the carbon emission factors for natural gas and purchased electricity, respectively. Similarly, to assess the relative environmental performance, the CO_2 Reduction Rate is defined:

$$CO_2 \text{ Reduction Rate} = \frac{ECM_{base} - ECM_{scenario}}{ECM_{base}} \times 100\% \tag{4.29}$$

where ECM_{base} and $ECM_{scenario}$ are the annual CO_2 emissions of the baseline and evaluated scenarios, respectively.

By employing this two-stage optimization framework and evaluating the outcomes using the defined performance metrics, we gain a clearer understanding of the benefits brought by integrating the Carnot battery into the industrial energy system. This will be illustrated in detail through the case study presented below. The optimization results yield a Pareto frontier, as shown in Figure 4.4, which intuitively depicts the balance between economic and environmental objectives.

4.3.4 Sensitivity analysis based on shadow price method

While the parameter-scan analysis discussed in Section 4.3.3 offers valuable insight into how the system's performance responds to large variations in external conditions, the shadow price approach allows for a finer examination of its economic sensitivity. Within the framework of a constrained optimization problem, the shadow price—or dual variable—of a given constraint indicates how much the optimal objective value would change if that constraint were relaxed by an infinitesimally small margin.

In our IES operation scheduling problem, the main goal is to minimize the overall operating cost. The shadow price corresponding to a specific operational constraint—for instance, the available renewable energy or the maximum storage capacity—indicates exactly how much the total operating cost would drop if that constraint were loosened by one unit. Such information serves as a strong economic indicator, helping to evaluate the marginal value of various resources and the cost impact imposed by system constraints.

The IES optimization problem is formulated as follows:

$$\min C(y) \tag{4.30}$$

s.t.

$$h(y) = 0 \tag{4.31}$$

$$g(y) \le b \tag{4.32}$$

where $C(y)$ is the total cost function, and the vectors $h(y)$ and $g(y)$ represent the equality (e.g., energy balance) and inequality (e.g., capacity limits) constraints, respectively, with b being the vector of constraint bounds. The shadow price, lambda, associated with a specific inequality constraint $g(y) \le b$ is defined as the partial derivative of the optimal cost function C^* with respect to the constraint bound b_i:

$$\lambda_i = \frac{\partial C^*}{\partial b_i} \tag{4.33}$$

When a resource constraint, such as available PV generation, shows a high positive shadow price, it means that adding one more unit of that resource could greatly reduce the total operating cost—making it particularly valuable. In contrast, if a capacity limit has a high shadow price, it points to that component being a key bottleneck, where capacity expansion could yield significant economic gains. This method will be applied in the following case study to offer deeper economic insights into how the system operates under different conditions.

4.4 CASE STUDIES AND ANALYSIS

To empirically validate the proposed two-stage optimization methodology and to quantitatively assess the multifaceted benefits of integrating a Carnot battery, this section presents a comprehensive case study. The selected site, the Ruili–Muse Cross-Border Industrial Park, was chosen for its complex and realistic energy profile, which serves as an ideal testbed for evaluating the performance of an advanced industrial IES. The park's diverse energy demands, coupled with its significant potential for renewable energy generation and waste heat recovery, create a rich analytical environment to probe the true value of sophisticated electro-thermal energy storage solutions.

4.4.1 Case background and data

Situated in Ruili City of Yunnan Province's Dehong Dai and Jingpo Autonomous Prefecture, the Ruili–Muse Cross-Border Industrial Park lies directly along China's border with Myanmar. It plays a central role in a national pilot zone for trade and industrial cooperation, acting as a regional hub for manufacturing and cross-border business. Its planned total area is 24 km^2, with an initial 5 km^2 phase already hosting over 40 enterprises in energy-intensive sectors such as papermaking and garment manufacturing. The predominant operational schedule for these businesses is a single-shift system, running daily from 8:00 AM to 12:00 AM.

Energy supply in the park is jointly supported by the regional power grid and a nearby natural gas gate station. Although the main utility pipelines have been constructed, individual enterprises handle their own end connections. The current IES is composed of conventional facilities—including CHP units, GBs, and EBs—to ensure reliable supply of both electricity and industrial heat.

To support the broader strategy of creating a cleaner and more resilient energy system, the park is actively expanding its renewable energy capacity. The area benefits from excellent natural conditions, with over 2000 h of annual sunlight and an average wind speed of 4–5 m/s, placing it within an efficient range for renewable power generation. These advantages lay a strong foundation for both solar and wind energy deployment. Nevertheless, the intermittent characteristics of these resources create substantial challenges for stable operation, especially given the industrial park's relatively inflexible and continuous energy demand.

Located in a tropical monsoon climate zone, Ruili experiences noticeable temperature variations across the year, which directly lead to large seasonal swings in the park's energy consumption. In summer, electricity demand surges due to the heavy use of air conditioning and industrial cooling systems, whereas in winter—despite the mild temperatures—there is a clear rise in energy consumption for space heating and process heat. This seasonal contrast underscores the need for a flexible energy system that can efficiently balance both electrical and thermal loads year-round. Figure 4.2 presents the normalized daily profiles of electricity and heat demand across the four seasons, vividly illustrating these dynamic variations. As depicted, each season exhibits a distinct energy consumption pattern: the summer curve (bottom left) shows a clear midday electricity peak with low and steady thermal demand, reflecting strong cooling requirements; in contrast, the winter curve (bottom right) demonstrates sustained high thermal demand, especially in the afternoon and evening, while electricity use remains relatively even. These patterns highlight the inherent complexity of electro-thermal coupling within the IES.

The local renewable resource profiles are visualized in the heatmaps of Figures 4.3 and 4.4. The solar resource, depicted in Figure 4.3, follows a predictable diurnal and seasonal pattern, with the highest energy output concentrated between 9:00 AM and 5:00 PM, and peaking during the summer months.

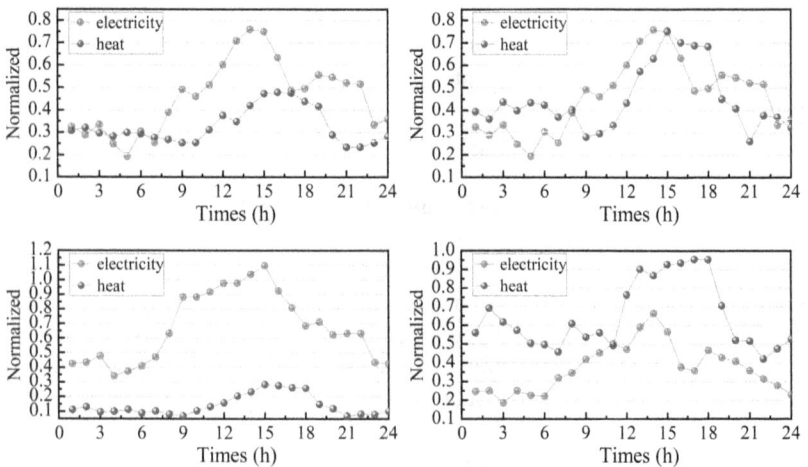

FIGURE 4.2 Normalized typical daily electrical and thermal load curves for four seasons.

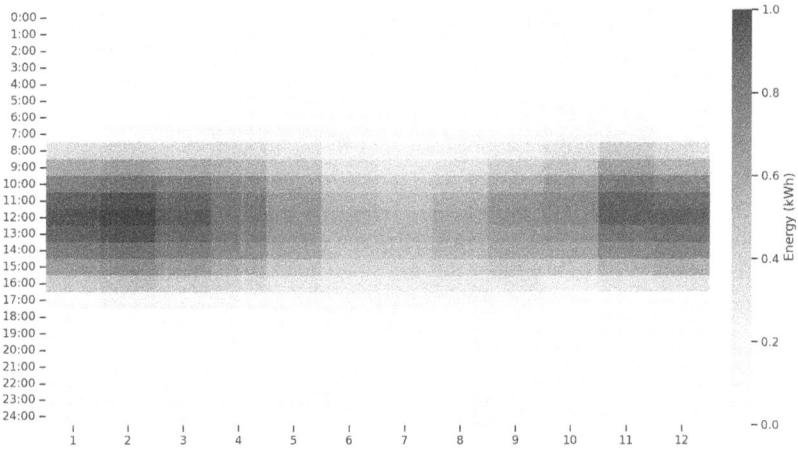

FIGURE 4.3   Heatmap of normalized annual PV power output.

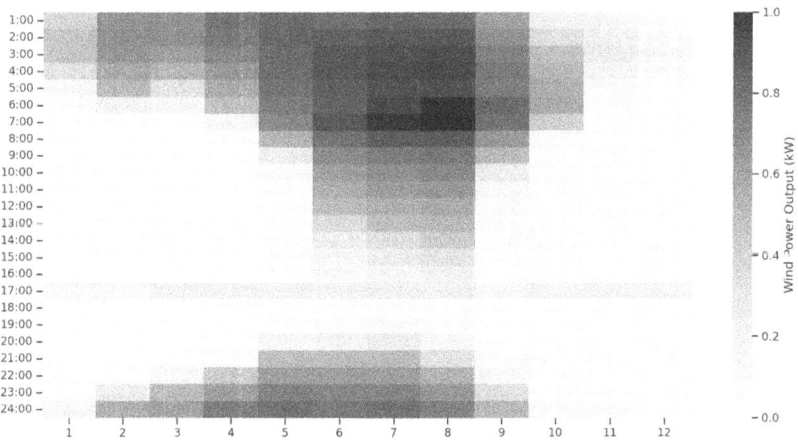

FIGURE 4.4   Heatmap of normalized annual WT power output.

This creates a temporal mismatch with the park's energy demand, which often extends late into the evening.

Conversely, the wind resource, shown in Figure 4.4, is more stochastic and complementary to solar. Significant wind power is often available during the night and in the early morning hours, as well as during seasons when solar irradiance is lower. This complementarity is valuable, but the inherent unpredictability of both resources necessitates a robust energy storage and dispatch strategy to ensure a reliable supply that matches the industrial park's operational schedule.

To incentivize efficient grid utilization, a time-of-use (TOU) electricity pricing policy is in effect, as detailed in Table 4.1. This price structure creates clear economic signals for load shifting and energy storage. The prices for other primary energy carriers are fixed, with natural gas priced at 3.89 RMB per cubic meter and hot water sold at 12.54 RMB per ton.

The techno-economic parameters of the energy equipment considered for optimization are provided in Table 4.2. These data form the critical inputs for calculating the investment, maintenance, and operating costs within the optimization model, thereby grounding the study in realistic financial and engineering constraints.

4.4.2 Scenario design and comparative analysis

To comprehensively assess how various storage configurations influence the performance of the park's IES, we developed four distinct scenarios. Each

TABLE 4.1 TOU electricity tariff.

PERIOD	TIME	PRICE (RMB/KWH)
Super peak	10:00–12:00, 17:00–19:00 (Mar–Apr, Dec–Jan)	0.13302
Peak	9:00–13:00, 17:00–22:00 (Daily)	0.12135
Flat	7:00–9:00, 13:00–17:00, 22:00–23:00 (Daily)	0.08039
Off-peak	23:00–7:00 (Daily)	0.04094

TABLE 4.2 Economic and technical parameters of energy system components.

ENERGY UNIT	INITIAL INVESTMENT (RMB/KW)	EFFICIENCY (%)	MAINTENANCE COEFFICIENT	SERVICE LIFE (YEARS)
PV	6000	—	0.02	25
Wind turbine	7200	—	0.024	22
CHP	4250	Elec: 33%, Heat: 50%	0.05	30
Gas boiler	850	85%	0.02	20
Electric boiler	600	95%	0.02	25
Battery	1800	95%	0.012	20
Carnot battery	3584	Elec: 10%, Heat: 90%	0.01	20

scenario follows the same two-stage optimization framework, differing mainly in the type of energy storage technology implemented.

This design aims to distinguish the individual role of each storage technology and to explore the possible synergies that emerge when they operate together within the system.

Scenario #1 serves as the baseline condition without any form of energy storage. This setup is essential because it mirrors the existing operational state—where the system depends solely on immediate self-generation and grid electricity to satisfy its demand. It also functions as a fundamental benchmark for comparing and assessing the performance of all other storage-integrated scenarios.

Scenario #2 incorporates only the electrochemical battery. The aim of this scenario is to assess the role and economic value of conventional electrical energy storage. Its main function is to perform peak shaving and valley filling by storing low-cost electricity during off-peak periods and discharging it during peak hours when prices are high. This configuration focuses on capturing potential profits purely from electricity price arbitrage.

Scenario #3 includes only the Carnot battery and plays a pivotal role in evaluating its distinctive advantages. In contrast to electrochemical batteries, the Carnot battery is capable of handling both electricity and heat flows at the same time, particularly excelling at capturing and making effective use of low-grade waste heat—an energy source that is otherwise left unused in the other scenarios.

Scenario #4 features a hybrid storage system incorporating both electrochemical batteries and a Carnot battery. This setup represents an integrated, advanced configuration that leverages the complementary advantages of the two technologies. The purpose is to examine whether their combined operation can generate synergistic effects greater than the sum of their standalone contributions.

In each scenario, the optimization model identifies the optimal capacity configuration for all generation and storage systems. The Pareto-optimal solution situated at the "knee point," where the trade-off between economic cost and carbon reduction achieves the best balance, is then chosen for detailed comparative analysis. The key performance indicators of the four scenarios are summarized in Table 4.3.

The results of the comparative analysis yield several critical and, in some cases, non-obvious insights. As expected, Scenario #1 establishes the baseline with the highest total annualized cost of 11.05 million RMB and the largest carbon footprint.

The introduction of an electrochemical battery in Scenario #2 produces a striking and counterintuitive result: a negative cost savings rate of −13.9%. While the battery does facilitate a respectable CO_2 reduction of 6.23% by

TABLE 4.3   Comparison of economic and carbon emission performance across scenarios.

METRIC	SCENARIO #1	SCENARIO #2	SCENARIO #3	SCENARIO #4
Annualized costs (RMB 10k)				
Initial investment	1482	1550	1704	1668
Maintenance	29	31	34	33
Electricity purchase	1238	1144	1370	1259
Gas purchase	2170	2033	1524	1618
Annualized total revenue (RMB 10k)	1529	1759	2043	2060
Annualized total cost (RMB 10k)	1105	1259	642	287
Cost savings Rate (%)	—	−13.9%	41.9%	74.0%
CO_2 emissions (tons)	577	541	542	559
CO_2 reduction rate (%)	—	6.23%	6.07%	3.12%

enabling greater absorption of renewable energy, the economic benefits derived from this electrical arbitrage are insufficient to offset the high capital and maintenance costs of the battery system. This finding strongly suggests that in an industrial context characterized by significant thermal loads, the value proposition of a standalone electrochemical battery, which is blind to the thermal dynamics of the system, may be limited and economically unviable.

In stark contrast, the integration of the Carnot battery in Scenario #3 fundamentally alters the system's economic performance. The total annualized cost plummets to 6.42 million RMB, yielding a substantial cost savings rate of 41.9%. This superior economic performance is directly attributable to the Carnot battery's unique capability to bridge the electrical and thermal domains. It not only stores surplus electricity but, crucially, valorizes the facility's low-grade waste heat, transforming a previously discarded by-product into a valuable energy input. This fundamentally changes the energy management paradigm from simple electrical storage to a holistic electro-thermal optimization.

Scenario #4, which combines both storage technologies, emerges as the optimal configuration from a purely economic perspective. It achieves the lowest total annualized cost of 2.87 million RMB, resulting in an extraordinary cost savings rate of 74.0%. This remarkable performance is not merely additive but is a clear demonstration of technological synergy. In this configuration, the Carnot battery serves as the system's workhorse for large-scale, long-duration energy storage and the strategic management of thermal resources, while the electrochemical battery provides high-power, fast-response services for short-term grid balancing and ancillary services. They perform distinct but complementary roles.

However, a closer examination of the environmental metrics reveals a crucial trade-off. While the hybrid system of Scenario #4 is the most economically advantageous, its CO_2 reduction rate (3.12%) is the lowest among the storage-inclusive scenarios. This is a direct consequence of the optimization's primary objective to minimize cost. In the hybrid system, the algorithm identifies that the most cost-effective operational strategy involves increasing the utilization of the high-efficiency CHP unit. While this tactical decision reduces reliance on more expensive grid electricity during peak hours, the corresponding increase in natural gas consumption leads to a slightly higher carbon footprint compared to scenarios where a different operational strategy might be favored. This highlights the inherent tension between pure economic optimization and environmental objectives, particularly in the absence of a strong carbon pricing signal that would financially penalize emissions more heavily.

4.4.3 Parameter sensitivity analysis

To assess the robustness of the proposed system and pinpoint the major external factors that influence its performance, a sensitivity analysis was carried out for Scenario #4, which features the hybrid energy storage configuration. This analysis explores how changes in electricity and natural gas prices, as well as in the availability of waste heat (within a range of −20% to +20%), impact the system's overall cost and emission levels.

4.4.3.1 Impact of electricity price fluctuations

As presented in Table 4.4, changes in electricity prices have a clear and immediate effect on the system's economic performance. A drop in electricity price reduces the ATC and significantly boosts cost savings. In contrast, when electricity prices rise, the total cost increases while the cost savings rate declines sharply, sometimes even turning negative. This pronounced sensitivity underscores the critical role of TOU pricing, since effective system optimization depends on purchasing low-cost electricity during off-peak hours for storage and on-site utilization.

TABLE 4.4   System performance under electricity price fluctuation.

METRIC (RMB 10K)	PRICE FLUCTUATION	−20%	−10%	0% (BASELINE)	+10%	+20%
Initial investment	1668	1668	1668	1668	1668	1668
Electricity cost	1259	1388	1295	1259	1160	1102
Gas cost	1618	1462	1574	1618	1779	1892
Total cost	287	−101	97	287	465	634
Cost savings (%)	—	135%	66%	74%	−62%	−121%
CO_2 (tons)	559	580	564	559	553	551
CO_2 Reduction (%)	—	−3.76%	−0.89%	3.12%	1.07%	1.43%

4.4.3.2 Impact of natural gas price fluctuations

As shown in Table 4.5, the system's economic outcomes are strongly influenced by fluctuations in natural gas prices, which determine the fuel costs for CHP units and GBs. When natural gas prices fall, the total cost drops noticeably; however, when prices rise, the overall cost increases accordingly. In addition, higher natural gas prices tend to cause a minor uptick in CO_2 emissions, since the system may shift toward greater use of grid electricity or other energy sources with varying carbon intensity.

4.4.3.3 Impact of waste heat fluctuations

The sensitivity analysis reveals a particularly remarkable insight: the system's performance is far more sensitive to waste heat availability than expected (Table 4.6). A 20% increase in recoverable waste heat leads to a 28% boost in cost savings, while a 20% reduction causes the savings rate to turn negative. This clear positive correlation demonstrates that waste heat should not be viewed as a mere by-product, but as a strategically valuable energy asset. The Carnot battery excels at capturing and converting this underutilized energy

TABLE 4.5   Impact of natural gas price fluctuations on system performance.

METRIC (RMB 10K)	PRICE FLUCTUATION	−20%	−10%	0% (BASELINE)	+10%	+20%
Initial investment	1668	1668	1668	1668	1668	1668
Electricity cost	1259	1100	1160	1259	1292	1374
Gas cost	1618	1896	1779	1618	1576	1477
Total cost	287	−69	115	287	447	602
Cost savings (%)	—	124%	60%	74%	−56%	−110%
CO_2 (tons)	559	551	553	559	564	578
CO_2 Reduction (%)	—	1.43%	1.07%	3.12%	−0.89%	−3.40%

TABLE 4.6   Impact of waste heat fluctuations on system performance.

METRIC (RMB 10K)	WASTE HEAT FLUCTUATION	−20%	−10%	0% (BASELINE)	+10%	+20%
Initial investment	1668	1668	1668	1668	1668	1668
Electricity cost	1259	1218	1233	1259	1293	1330
Gas cost	1618	1775	1699	1618	1531	1440
Total cost	287	380	330	287	246	206
Cost savings (%)	—	−32%	−15%	74%	14%	28%
CO_2 (tons)	559	564	561	559	558	558
CO_2 Reduction (%)	—	−0.89%	−0.36%	3.12%	0.18%	0.18%

stream into economic value, fundamentally changing how IESs are evaluated. Acting as an efficient converter, it transforms low-cost waste heat into high-value electricity and thermal energy—solidifying its role as a key enabler of circular economy principles in industrial applications.

4.5 CONCLUSION AND FUTURE OUTLOOK

The analysis presented in this chapter culminates in a clear and compelling argument for the strategic integration of electro-thermal energy storage, particularly the Carnot battery, as a cornerstone technology for the decarbonization and economic optimization of industrial IESs. By establishing a rigorous two-stage optimization framework and applying it to a realistic case study, this work has moved beyond theoretical modeling to provide quantitative evidence of the profound impact that different storage configurations have on system performance. The findings not only highlight the immense potential of the Carnot battery but also expose the limitations of conventional, single-vector energy storage solutions in the complex, multi-energy environment of modern industry.

The comparative scenario analysis yielded several critical insights. First, it demonstrated unequivocally that in an industrial context with significant thermal loads, the economic viability of standalone electrochemical batteries is highly questionable. While effective at reducing emissions through renewable energy arbitrage, the high capital costs were not offset by the operational savings from purely electrical peak shaving, leading to a negative return on investment. This counterintuitive result serves as a crucial qualifier to the often-generalized benefits of battery storage, emphasizing that a technology's value is deeply context-dependent.

In stark contrast, the Carnot battery emerged as a transformative technology. Its ability to valorize low-grade waste heat—transforming an otherwise discarded by-product into a strategic energy input—fundamentally altered the system's economic calculus. This capability for electro-thermal synergy is the Carnot battery's defining advantage, allowing it to generate a substantial 41.9% CSR even as a standalone unit. The subsequent analysis of a hybrid system, combining the Carnot battery with an electrochemical battery, revealed the potential for even greater synergy, achieving a remarkable 74.0% cost reduction. In this optimal configuration, the two technologies perform distinct but complementary roles: the Carnot battery acts as the system's workhorse for large-scale, long-duration energy management and thermal optimization, while the electrochemical battery provides high-power, fast-response services for grid stability and ancillary functions.

However, this research also illuminated the inherent tension between purely economic optimization and the pursuit of environmental objectives. The discovery that the most economically advantageous hybrid system yielded a lower CO_2 reduction rate than other storage scenarios is a critical finding. It underscores the reality that, in the absence of a robust carbon pricing mechanism, the most profitable operational strategy may not align with the most environmentally benign one. This highlights a significant policy gap and suggests that achieving deep decarbonization in the industrial sector will require not only technological innovation but also market signals that properly price carbon emissions, thereby aligning financial incentives with climate goals.

Looking ahead, the findings from this study open up several promising avenues for future research. While this work provides a robust foundation, further advancements are needed to accelerate the real-world deployment and optimization of Carnot batteries in industrial settings. Three key areas warrant particular attention:

1. Development of high-fidelity, computationally efficient models: The thermodynamic models used in this study, while accurate, are computationally intensive. This poses a significant challenge for their application in large-scale, multi-year planning optimizations or in real-time control systems. Future work should focus on developing high-fidelity surrogate models, likely based on artificial intelligence techniques such as neural networks. Such models could drastically reduce the computational burden of simulating Carnot battery performance across a wide range of operating conditions, making it feasible to embed them within more complex dynamic optimization and model-predictive control (MPC) frameworks.

2. Exploration of advanced system configurations and working fluids: The Carnot battery configuration analyzed in this chapter represents

a foundational design. There is significant scope for innovation in its system architecture and material selection. Future research should investigate alternative thermodynamic cycles, such as those incorporating multi-stage compression with intercooling or recuperation, to further enhance RTE and COP. Furthermore, a systematic exploration of different working fluids and thermal storage media is needed to tailor Carnot battery designs to the specific temperature ranges and chemical compatibility requirements of diverse industrial waste heat streams.

3. Incorporation of dynamic scheduling and uncertainty: This study relied on representative daily profiles for load and renewable generation. Real-world industrial operations, however, are subject to significant stochasticity and unforeseen fluctuations. The next generation of research must move toward dynamic operational scheduling. This involves developing sophisticated control strategies, such as MPC, that can respond in real time to variations in energy prices, renewable energy availability, and industrial process demands. Such dynamic frameworks would enable the IES to continuously optimize its performance under uncertainty, maximizing both economic returns and environmental benefits in a constantly changing operational landscape.

In conclusion, the Carnot battery is not merely an alternative to electrochemical storage; it is a fundamentally different class of technology that offers a holistic solution to the coupled electro-thermal challenges of the industrial sector. By unlocking the value of waste heat and providing flexible, multi-vector energy services, it has the potential to be a key enabler of a more efficient, resilient, and sustainable industrial future. The path forward lies in refining the technology, developing smarter control strategies, and implementing supportive policies that recognize its unique and powerful capabilities.

Conclusion and Future Outlook

5

Yihui Mao, Xiaojie Lin, Wei Zhong, and Jian Song

5.1 CONCLUSION

This monograph has traversed the landscape of Carnot battery technology, from its foundational thermodynamic principles to its complex integration within modern industrial energy systems. The preceding chapters have systematically constructed a narrative that positions the Carnot battery not merely as an alternative energy storage device, but as a pivotal enabling technology for a more efficient, flexible, and sustainable energy future. The final chapter of this study synthesizes the key findings, explores the broader implications of the technology, and charts a course for future research and development. This research and development will be critical for the technology's widespread commercial deployment.

The investigation commenced in Chapter 1 by establishing the imperative for long-duration energy storage (LDES) in an era dominated by the variable output of renewable energy sources. The Carnot battery, alternatively designated as pumped thermal electricity storage, was introduced as a thermodynamically sophisticated solution that facilitates the conversion of electricity

DOI: 10.1201/9781003630821-5

into thermal potential and vice versa. It has been determined that the core value proposition of the entity in question is defined by three distinguishing attributes: (1) Duration scalability: The utilization of plentiful and inexpensive geological materials, such as rocks or molten salts, for thermal storage enables the cost-effective extension of storage duration to tens of hours, thereby overcoming a primary economic barrier associated with electrochemical systems. (2) The sector-coupling capabilities of the aforementioned technologies are of particular relevance. The inherent ability to provide electricity, heating, and cooling from a single installation creates diversified revenue streams and enhances overall system efficiency, positioning it as a key technology for integrating energy sectors. (3) The issue of siting flexibility must be addressed. In contrast to geographically constrained LDES technologies, such as pumped hydro, Carnot Batteries can be strategically deployed in proximity to load centres or renewable generation hubs. This approach minimizes infrastructure requirements and optimizes grid benefits.

Chapter 2 provides a comprehensive examination of the design, modelling, and optimization of Carnot battery systems. The text delineated two primary system architectures – Rankine-based and Brayton-based cycles – each with distinct operational characteristics suited to different temperature ranges and applications. A key finding was the multifaceted nature of component selection, where the choice of working fluids, thermal storage media (sensible, latent, or thermochemical), and turbomachinery directly dictates system performance, cost, and safety. The chapter also presented a detailed case study that employed parameter sensitivity analysis to identify heat source temperature, temperature difference, and heat pump temperature rise as key variables significantly impacting system efficiency. This underscored the necessity of sophisticated thermodynamic modelling – spanning steady-state, dynamic, and economic frameworks – to navigate the complex trade-offs inherent in system design and optimization.

As the discussion progresses from the component level to the system level, Chapter 3 provides a detailed exposition of the modelling of integrated energy system (IES), the complex environment into which Carnot batteries are deployed. The chapter contrasted holistic (energy hub) and component-based (energy bus) modelling approaches, establishing the latter as a more granular method for capturing the intricate interactions between diverse energy assets. The chapter under scrutiny emphasized the profound impact of time granularity on system scheduling, stemming from the different dynamic characteristics of electricity, gas, and thermal energy carriers. This necessitated the development of multi-scale scheduling strategies, such as model predictive control (MPC), which employ a rolling optimization approach to manage system dynamics and uncertainties more effectively than traditional static optimization algorithms.

Finally, Chapter 4 provided a comprehensive validation of the Carnot battery's value proposition through a case study of its integration into a real-world industrial IES at the Ruili-Muse cross-border industrial park. Employing a novel two-stage capacity configuration optimization method, which was solved with the NSGA-II algorithm, the analysis compared four scenarios with different storage configurations. The findings indicate that a standalone elec-trochemical battery has been deemed economically unviable within this indus-trial context, yielding a negative cost savings rate due to the high capital costs that exceed the benefits of electricity arbitrage alone. The introduction of a Carnot battery (Scenario #3) resulted in a significant shift in the economic calculus, achieving a cost savings rate of 41.9% through the effective utiliza-tion of low-grade industrial waste heat. In the fourth scenario, a hybrid system was employed, integrating the Carnot battery for bulk thermal and electrical storage with an electrochemical battery for rapid electrical balancing. This configuration yielded the optimal economic performance, with a noteworthy 74.0% cost reduction rate. This outcome underscores the synergistic benefits of hybrid storage solutions. It was found that the system's performance was dis-proportionately influenced by the availability of waste heat, as indicated by the sensitivity analysis. This significant finding redefines industrial waste heat, which is typically regarded as a thermodynamic by-product and thus destined for disposal, as a valuable strategic resource. The Carnot battery emerges as the pivotal technology that facilitates the economic exploitation of this resource.

The research results of this monograph are of great significance in the fol-lowing respects: Firstly, the Carnot battery is a significant catalyst for the estab-lishment of a circular energy economy. The process involves the capture and upgrading of low-grade industrial waste heat, thereby transforming a linear energy flow (where heat is produced, used, and discarded) into a circular one (where thermal energy is recovered, stored, and redeployed to generate elec-tricity or provide high-grade process heat). This capability is pivotal to the pro-cess of industrial decarbonization, providing a feasible route for energy-intensive sectors such as manufacturing and chemical processing to concurrently reduce energy expenditures, minimize emissions, and decrease reliance on fossil fuels for process heating. Secondly, the technology provides a pragmatic solution for infrastructure repurposing and a just transition. The decommissioning of fossil fuel power stations can be repurposed to create zero-emission energy storage facilities. This is achieved by replacing the boilers with thermal storage sys-tems, while retaining the valuable grid infrastructure, including interconnec-tions and rotating machinery. This approach has been demonstrated to have a number of key benefits. Firstly, it has been shown to preserve grid stability services such as inertia. Secondly, it has been demonstrated to provide ongoing employment for a skilled workforce. This, in turn, has the effect of mitigating the socioeconomic disruption of the energy transition in affected communities.

Thirdly, the hybrid storage configuration identified as optimal in Chapter 4 demonstrates a crucial principle: the future of energy storage is not a zero-sum competition between technologies, but rather a synergistic integration of complementary solutions. Electrochemical batteries offer unparalleled rapid response, while Carnot batteries provide cost-effective, long-duration bulk storage and thermal management. The integration of these technologies enables an IES to address the entire range of grid requirements, from millisecond-level frequency regulation to multi-day energy shifting.

5.2 FUTURE OUTLOOK

While this monograph has established the significant potential of Carnot batteries, continued research and development are essential to accelerate their journey from demonstration to widespread commercial adoption. Future work should focus on several key areas:

1. Research into novel thermal storage media, including more stable phase-change materials and higher-density thermochemical storage systems, could further improve energy density and performance. Concurrently, the development of efficient, reliable, and cost-effective reversible turbomachinery that can function as both a compressor and expander would significantly reduce system complexity and capital costs.

2. The computational demands of dynamic, system-level modelling present a challenge for real-time optimization. Future research could leverage artificial intelligence and machine learning to create high-fidelity surrogate models that enable faster and more sophisticated control strategies, such as advanced MPC, capable of optimizing system performance amid the real-time fluctuations of renewable generation and energy prices.

3. Further studies should explore the integration of Carnot batteries with a wider range of external heat sources, such as concentrated solar power and geothermal energy. Expanding the scope of application to urban energy systems, where the technology can provide simultaneous electricity storage and support for district heating and cooling networks, represents a significant growth opportunity.

4. The full value of Carnot batteries, particularly their contributions to grid stability, long-duration capacity provision, and sector coupling, is often not fully compensated in current electricity market

structures. Policy development and market reforms are needed to create frameworks that properly recognize and reward the diverse services these systems provide, thereby de-risking investment and levelling the playing field with incumbent technologies.

The transition to a global energy system powered by renewable resources is inevitable and will be achieved through a combination of technological and policy developments. The primary challenges are the management of intermittent energy sources, the integration of diverse energy sectors, and the optimization of energy generation efficiency. As demonstrated in this monograph, the Carnot battery is a uniquely versatile and powerful tool designed for this very challenge. The technology under discussion successfully bridges the gap between electrical and thermal energy, between short- and long-duration storage, and between energy generation and industrial processes. It offers a robust, scalable, and economically compelling pathway towards a decarbonized energy future. Its evolution from a thermodynamic concept to a commercially viable technology is indicative of ongoing innovation and suggests that solutions to the most complex energy problems may be found in the elegant and enduring principles of thermodynamics.

Index

Pages in *italics* refer to figures and pages in **bold** refer to tables.

For Product Safety Concerns and Information please contact our EU
representative GPSR@taylorandfrancis.com
Taylor & Francis Verlag GmbH, Kaufingerstraße 24, 80331 München, Germany

* 9 7 8 1 0 4 1 0 4 9 8 0 7 *